T0139960

Intelligent Systems Reference Library

Volume 78

Series editors

Janusz Kacprzyk, Polish Academy of Sciences, Warsaw, Poland
e-mail: kacprzyk@ibspan.waw.pl

Lakhmi C. Jain, University of Canberra, Canberra, Australia, and
University of South Australia, Adelaide, Australia
e-mail: Lakhmi.Jain@unisa.edu.au

Konstantina Chrysafiadi · Maria Virvou

Advances in Personalized Web-Based Education

 Springer

Konstantina Chrysafiadi
Maria Virvou
Department of Informatics
University of Piraeus
Piraeus
Greece

ISSN 1868-4394   ISSN 1868-4408   (electronic)
ISBN 978-3-319-36048-5   ISBN 978-3-319-12895-5   (eBook)
DOI 10.1007/978-3-319-12895-5

Springer Cham Heidelberg New York Dordrecht London

Printed on acid-free paper

Springer is part of Springer Science+Business Media (www.springer.com)

Preface

This book aims to provide important information to researchers, educators, and software developers of computer-based educational software ranging from e-learning and mobile learning systems to educational games, including stand-alone educational applications and intelligent tutoring systems. In particular, this book explains how fuzzy logic can be used to automatically model the learning or forgetting process of a student. Also, it describes an innovative module, which is responsible for tracking cognitive state transitions of learners with respect to their progress or non-progress. Therefore, this book shows how personalized tutoring modeling may be achieved by taking into account either how a student is making progress in learning or how the student's knowledge can decrease. In order to make the student modeling process clear, a review of the literature concerning student modeling during the past decade is included in a special chapter. This chapter aims to answer the three basic questions on student modeling: what to model, how and why. It presents comparative tables that are the results of a 10-year review study in student modeling. So the particular chapter can be, also, used as a guide for making decisions about the techniques that should be adopted when designing a student model for an adaptive tutoring system. However, the work presented in this book is not limited to adaptive instruction, but can also be used in other systems with changeable user states, such as e-shops, where consumers' preferences change over time and affect one another. Thereby, this book can provide important information not only to those interested in educational systems and student modeling, but also to all researchers and software developers who are interested in user modeling in any adaptive and/or personalized system.

Contents

Abstract

The rapid advances of computer technology and Internet have led to an enormous growth of interest in the field of e-learning applications. However, e-learning systems have several shortcomings, which concern adaptivity problems, when compared with real-classroom education. Therefore, this book aims to provide important information about adaptivity in computer-based and/or web-based educational systems. Initially, a literature review on student modeling techniques and approaches during the past decade is presented. Then, a novel student modeling approach, which maximizes the effectiveness of learning and adaptivity, is presented.

This book presents how fuzzy logic can be used for offering adaptation and increasing learning effectiveness in Intelligent Tutoring Systems. In particular, it presents a hybrid student model, which incorporates a rule-based mechanism that allows each individual learner to complete the training program in her/his own learning pace and abilities. The presented student model combines an overlay model and stereotypes with fuzzy sets and fuzzy rules. It is responsible for identifying and updating the learner's knowledge level for all the domain concepts of the learning material each time. Particularly, each time the learner's knowledge level on a domain concept changes, the system has to infer how the learner's knowledge level on the related concepts also changes. In this way, the system discovers if the student learns or not, if s/he forgets, if s/he has difficulties in understanding, if s/he assimilates the knowledge. Therefore, the presented approach models either how learning progresses or how the student's knowledge can be decreased. As a result, the system adapts the delivery of the learning material to each individual learner's need and pace. The operation of the presented approach is based on a Fuzzy Network of Related-Concepts (FNR-C), which is a combination of a network of concepts and fuzzy logic. It is used to represent the organization and structure of the learning material as the knowledge dependencies that exist between the domain concepts of the learning material.

The presented novel approach was fully implemented and evaluated. It was integrated in a programming tutoring system for the programming language 'C'. Students of a postgraduate program in the field of Informatics on the University of Piraeus, Greece, used the particular system to learn how to program with the programming language 'C'. The evaluation results were very encouraging. They demonstrated that the presented student modeling approach had a positive impact on the learners' performance and on the learning process. Furthermore, they showed that the system made valid and meaningful adaptation decisions. The gain of the presented approach is that it allows the system to model the student's knowledge level and the learning process in a more realistic way. Furthermore, the particular approach constitutes a novel generic tool, which is able to model the changeable use states (e.g., knowledge level, preferences, emotions).

Introduction

During the last decades there has been an ever-growing increased interest in e-learning applications. The reason is either the easy access of e-learning applications by a large and heterogeneous group of learners at any time and place, or the challenge to develop an adaptive e-learning system. The goal of each web-based educational system is to maximize the effectiveness of learning and introduce the learning and teaching process of real-classroom education to the web. However, in real classrooms, human teachers can readjust the instructional process and their teaching strategy, each time they think that the learning outcomes of their students fall short of their teaching expectation. Consequently, the challenge is to develop Web-based educational systems that adapt dynamically to each individual student for effective delivery of knowledge domain to heterogeneous student populations.

Learners of web-based educational systems have not only different needs, but also different learning characteristics. They have different knowledge level, cognitive and meta-cognitive abilities, preferences, learning styles, emotions, reactions, etc. An adaptive educational system has to identify each individual student's needs and learning characteristics and react accordingly offering effective personalization. Therefore, an adaptive educational system has to identify the student's learning characteristics, to infer her/his needs and preferences, to deliver the appropriate learning material adapted to the student's needs, to advise the learner and to provide personalized feedback. In this way, the system facilitates and maximizes the effectiveness of the learning process.

Each learner has her/his own learning pace, and consequently educational environments have to adapt to this. In fact, it is pedagogically ineffective to deliver the same learning material and provide the same instructional conditions to all the learners without considering their learning needs and characteristics. Not all learners should be told to read the same material in the same order. Instead, learning material should be delivered with respect to students' knowledge level and personal needs. Furthermore, the developers of personalized and/or adaptive educational systems have to consider that the learner's knowledge of a domain concept is subject to change. Hence web-based educational systems have to rec-ognize the individual learner's knowledge level and how this changes, and then they should provide effective instruction, adapting the delivery of the knowledge domain to the learner's learning needs and pace.

A solution to the above problem is the technology of Intelligent Tutoring Systems (ITSs). ITSs are computer-based tutoring systems, which incorporate arti-ficial intelligence and thus they can adapt dynamically the content and instruction to the individual student's needs and preferences offering a highly personalized learning experience. The success of adaptation is based on the four modules of the Intelligent Tutoring System (ITS): the interface module; the knowledge domain module; the student model; and the tutoring module.

The knowledge domain representation is an important aspect that has to be specified for offering adaptation. The knowledge domain representation is a description of expertise in the subject-matter domain of the ITS. The most popular techniques used for knowledge domain representation are: hierarchies, network of concepts, linkage graphs, and concept maps. These techniques are used either to represent the order in which each domain concept has to be taught or the knowledge dependencies that exist between the domain concepts of the learning material. However, the representation of the relations between concepts is, mainly, restricted to "part-of", "is-a", and prerequisite relations. Yet, there is the need to represent how the student's knowledge level on a domain concept is affected by her/his knowledge level on other related domain concepts. In such cases, the rep-resentation of this kind of relations of the learning material's domain concepts is performed using fuzzy techniques. The combination of a network of concepts and fuzzy logic creates a Fuzzy Related-Concepts Network (FR-CN). FR-CN is a network of concepts, which also depicts the knowledge dependencies that exist between the domain concepts of the learning material. However, the knowledge domain representation has to be combined with a well-designed student model, which is responsible for how the system uses the knowledge domain module to make the right decisions to offer personalized instruction and support to the learner.

The ability of an ITS to provide adaptivity is based, mainly, on the technology of student modeling (Devedzic 2006). Student modeling has been introduced in ITSs, but its use has been extended to most current educational software appli-cations that aim to be adaptive and personalized. Student modeling allows the system to identify the students' needs and leads it to make adaptive instructional decisions. This means that the system generates hypotheses about students' needs based on evidence that has been previously collected silently during the learner's interaction with the system. In return, it provides personalized tutoring to each individual student.

There are a variety of techniques for student modeling. The most widely known techniques are overlay models and user stereotypes. Other techniques for student modeling are: perturbation, machine learning techniques, cognitive theories, con-straint-based model, fuzzy logic techniques, Bayesian networks, ontologies. Each student modeling technique is appropriate for modeling some particular students' characteristics. For example, the overlay model is, usually, used for representing the student's knowledge level; stereotypes are preferred to model the student's learning styles and preferences; cognitive theories are used for modeling the affective features of students, etc. Many researchers have used a combination of the above techniques to model more than one features of the students.

Frequently, student modeling deals with uncertainty. Learning is a complicated process. It cannot be accurately hypothesized that a learner knows or does not know a domain concept. For example, a new domain concept may be completely unknown to the learner but in other circumstances it may be partly known due to previous related knowledge of the learner. On the other hand, domain concepts, which were previously known by the learner, may be completely or partly forgotten. Hence, currently they may be partly known or completely unknown. In this sense, the level of knowing cannot be accurately represented. Finally, the teaching process itself changes the status of knowledge of a user. This happens due to the fact that a student learns new concepts while being taught.

In view of the above, the representation of the learner's knowledge is a moving target. A solution to this problem is the use of fuzzy logic. Fuzzy logic allows the system to model either the increase or the decrease on the student's knowledge level. In particular, fuzzy logic techniques can be used to model how the learner's knowledge level on a domain concept of the learning material is affected by changes in her/his knowledge level on another related concept. This means that fuzzy logic technique can model the uncertain and inaccurate states of learning and forgetting.

In view of the above, a novel approach in ITS, which includes fuzzy logic techniques, is presented in the particular book. More specifically, it includes a rule-based fuzzy logic mechanism in combination with an overlay model and user stereotypes for providing personalized tutoring to each learner. This mechanism identifies either the domain concepts that the learner has forgot or the concepts that s/he has learned. Therefore, the presented fuzzy student model reveals if a student learns or not, if s/he forgets or if s/he assimilates the learning material and allows the system to adapt the instruction to each individual student's learning pace.

The presented novel fuzzy system was fully implemented in a web-based programming tutoring system that teaches the programming language 'C'. The reason for the selection of the particular knowledge domain is the fact that the need for adaptivity is crucial in the programming tutoring system. In the domain of computer programming, there are many different programming languages and learners have different backgrounds and characteristics.

Programming language learners can vary from novice programmers, to more experienced programmers who know programming languages other than that being taught. Obviously, while learning a new programming language a novice programmer has to learn many more domain concepts than does a more experienced programmer, who already knows the principles and the basic structures of computer programming. Furthermore, if a learner already knows an algorithm (e.g., calculating the sum of integers in a 'for' loop), there is no need for her/him to learn another similar algorithm (e.g., counting in a 'for' loop). Similarly, if a learner knows a programming structure (e.g., one-dimensional arrays), it is easier to understand another programming structure (e.g., multidimensional arrays), so this new structure should not be considered as being completely unknown to the learner. On another occasion, if a learner's performance on a domain concept is poor, this suggests that she/he has forgotten another relevant domain concept. For example, if a learner has difficulties in calculating a sum in a 'while' loop, her/his knowledge of the previous domain concept of "calculating a sum in a 'for' loop" has eroded. In view of these problems, the presented web-based programming tutoring system incorporates a student model responsible for identifying and updating the student's knowledge level, taking the different pace of learning of each individual learner into account.

In particular, the presented fuzzy system retains static information about each student, such as her/his previous experience on computer programming and the programming languages that she/he already knows. It also retains dynamic information such as errors, misconceptions, and progress. Such kind of information is gathered during the learner's interaction with the system. In each learning session, the system recognizes the learner's knowledge level and the changes that occur in the state of her/his knowledge of a domain concept; it then updates the student's overall knowledge level according to the knowledge dependencies between the learning material's domain concepts and the learner's progress. The system recognizes when a new domain concept is completely unknown to the learner, or when it is partly known due to the learner having previous related knowledge. Furthermore, it recognizes when a previously known domain concept has been completely or partly forgotten by the learner. Thus it models either the possible increase or decrease of the learner's knowledge. Furthermore, each time it checks if the learner's errors were due to possible confusion with features of another previously known programming language. In this case, the system responds accordingly by adapting instantly the sequence of learning lessons. The personalization achieved, allows every learner to complete the e-training course on their own pace and ability.

The presented programming tutoring system was used by the students of a postgraduate program in the field of Informatics in the University of Piraeus, Greece, in order to learn how to program in programming language C. For the evaluation of the fuzzy student model approach, the evaluation framework PERSIVA (Chrysafiadi and Virvou 2013), which includes both questionnaires and observations through experiments, was used. The evaluation method assessed either the educational impact (i.e., performance, satisfaction, change of learners' attitudes) or the effectiveness and validity of the educational system's adaptivity are assessed. The results of the evaluation were very encouraging. They demonstrated that the system is able to adapt dynamically to each individual learner's needs by scheduling the sequence of lessons instantly. This personalization allows each learner to complete the e-training course at their own pace and according to their ability.

The main body of this book is organized into four chapters. The first chapter concerns a literature review of techniques and applications of student modeling for personalized education. The second chapter presents an overview of fuzzy logic and describes how fuzzy logic can be used for representation of the knowledge domain and in student modeling. In that chapter, the mechanism of fuzzy rules is described. The third chapter presents a novel hybrid student model for personalized education that the book's authors have created. That student model is responsible for identifying the improvements and the decay of the learner's knowledge. Furthermore, in that chapter, the implementation of the presented hybrid student model in an integrated programming tutoring system for the programming language 'C' is described. In the last chapter, the evaluation of the novel system is presented and discussed. Finally, the conclusions drawn from this work are presented.

Chapter 1
Student Modeling for Personalized Education: A Review of the Literature

Abstract The rapid development of computer technology and e-learning reinforces the need of dynamic adaptation to the needs of each individual student. Adaptation is performed through the student model, which is a crucial module of an Intelligent Tutoring System. There are many student modeling techniques and approaches. In this chapter, a review of the literature concerning student modeling during the past decade is presented. The aim is to answer the three basic questions on student modeling: what to model, how and why. This chapter presents comparative tables that are the results of a 10-year review study in student modeling. They reveal either the most common modeled student's characteristic, or the student modeling approaches that are preferred in relation to student modeling characteristics. So, the particular chapter can be, also, used as a guide for making decisions about the techniques that should be adopted when designing a student model for an adaptive tutoring system.

1.1 Introduction

"Intelligent Tutoring Systems" (ITSs) are computer-based educational systems that contain some intelligence and can be used for adaptive learning. Their goal is to maximize the effectiveness of e-learning as human tutors provide the most effective instruction at classroom level readjusting each time the instructional process and the teaching strategy considering the student's needs and abilities. The design and development of an ITS is based on techniques that combine theories and models from the computer science, cognitive science, psychology, learning science, computational linguistics, artificial intelligence (Nwana 1990; Graesser et al. 2012) (Fig. 1.1). The typical architecture of an ITS includes the following four modules (Fig. 1.2):

- A knowledge domain model that stores the learning material that is taught to students.
- A student model that stores information about the learner's knowledge level, abilities, preferences and needs.

© Springer International Publishing Switzerland 2015

K. Chrysafiadi and M. Virvou, *Advances in Personalized Web-Based Education*,
Intelligent Systems Reference Library 78, DOI 10.1007/978-3-319-12895-5_1

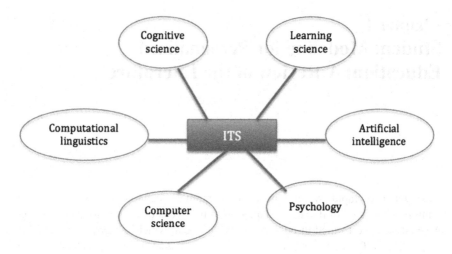

Fig. 1.1 An intelligent tutoring system

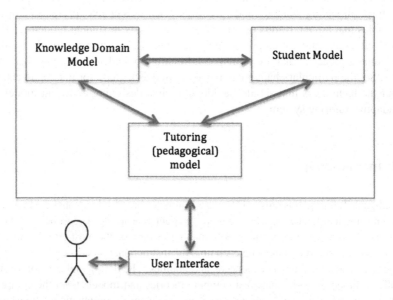

Fig. 1.2 The typical architecture of an ITS

- A tutoring (pedagogical) model, which makes student diagnosis and controls the tutoring process and make appropriate instructional decisions based on the information provided by the other components of the ITS.
- A User Interface that allows the system to interact with the user—learner.

The main feature and advantage of an ITS is its ability to adapt the content and the presentation of the learning material, the feedback and the instruction process and strategy to the student's needs and preferences. To fulfill this, artificial intelligence techniques are applied to the main components of the ITS' architecture. In other words, artificial intelligence techniques are applied to the knowledge domain representation, to the student model and to the pedagogical model (Badaracco and Martinez 2013). Researches in ITS include researches in techniques that will make an ITS to 'behave' in a more intelligent way (Conati 2009), which means to diagnose the student's learning status and needs in a more effective way and manage the instructional and pedagogical strategy as a real domain expert. Many of these researches have been extended to researches in student model. This has happened due to the fact that the student model is the base for personalization in computer-based educational applications.

As a consequence, a crucial factor for designing an effective ITS and/or an adaptive educational system is the construction of an effective student model. In order to construct a student model, it has to be considered what information and data about a student should be gathered, how it will update in order to keep it up-to-date, and how it will be used in order to provide adaptation (Millán et al. 2010; Nguyen and Do 2009). In fact, when a student model is constructed, the following three questions have to be answered:

(i)   What are the characteristics of the user we want to model?
(ii) How we model them?
(iii) How we use the user model?

The target of this chapter is to present the student's characteristics that are usually modeled. Furthermore, the student modeling techniques that are used in the literature in relation to each student's characteristic are presented.

1.2 Student Modeling Techniques and Methods

1.2.1 The Overlay Method

One of the most popular and common used student models is the overlay model. It was invented by Stansfield et al. (1976) and has been used in many systems ever since. The reason for its extensive use is the fact that the overlay model can represent independently the user knowledge for each concept. According to the overlay modeling, the student model is a subset of the domain model (Martins et al. 2008; Vélez et al. 2008), which reflects the expert-level knowledge of the subject (Brusilovsky and Millán 2007; Liu and Wang 2007) (Fig. 1.3). Therefore, the student's knowledge is represented as incomplete but no as incorrect. The incomplete student's knowledge is defined by the differences between her/his and the expert's set of knowledge (Bontcheva and Wilks 2005; Michaud and McCoy 2004; Staff 2001; Nguyen and Do 2008). According to the overlay student modeling approach, the knowledge domain is decomposed into individual topics and concepts that are called elements. Usually, each element is characterized as known or unknown for the student. However, there are overlay models, in which each element is

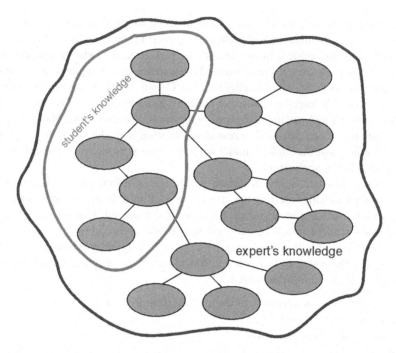

Fig. 1.3 Overlay model

characterized by a qualitative measure (good-average-poor) or a qualitative measure such as the probability that the student knows the concept (Brusilovsky and Millán 2007). This kind of representation informs the system about the degree to which the learner knows each domain element.

1.2.2 User Stereotypes

Stereotypes were introduced to user modeling by Rich (1979). The main idea of stereotyping is to create groups of students with common characteristics. Such groups are called stereotypes. A learner will be assigned into a related stereotype if some of his/her characteristics match the ones contained in the stereotype. For example, a stereotype model can present the knowledge lever (Fig. 1.4) or the learning style of a student. In these cases the stereotypes could be {novice, beginner, knowledgeable, advanced, expert} and {visual, verbal} accordingly. Each stereotypes has a set of trigger conditions, which activate the stereotype if they are true, and a set of retraction conditions, which deactivate the stereotype if they are true to Kay (2000). The stereotype is a particularly important form of reasoning about users and also student modeling with stereotypes is often a solution for the problem of initializing the student model by assigning the student to a certain group of

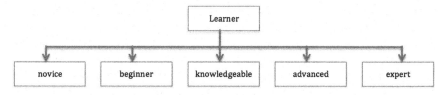

Fig. 1.4 Stereotypes of the learner's knowledge level

students (Tsiriga and Virvou 2002). An appealing property of the stereotype is that it should enable a system to get started quickly on its customized interaction with the user Kay (2000). However, stereotype approach is quite inflexible and error-prone due to the fact that stereotypes are constructed in a hand-crafted way before real users have interacted with the system and they are updated only by the system's designer or developer (Kass 1991; Tsiriga and Virvou 2002).

1.2.3 Models for Misconceptions and Erroneous Knowledge

1.2.3.1 Perturbation

A perturbation student model is an extension of the overlay model. It represents student's knowledge as a subset of the expert's knowledge along with her/his misconceptions (Mayo 2001; Nguyen and Do 2008) (Fig. 1.5). The perturbation student model is useful for diagnostic reasoning. It allows the system to identify the student's erroneous knowledge and wrong rules that s/he has applied and led her/him to wrong answer (Martins et al. 2008). Thus, the system remediate the student's misconceptions providing her/him the appropriate learning material, advices and feedback.

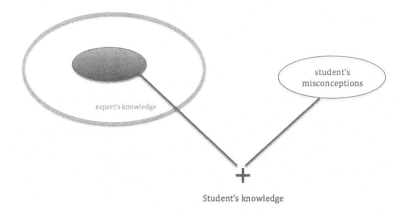

Fig. 1.5 Perturbation model

The perturbation model has a collection of mistakes, which is usually called bug library. The bug library and can be built either by empirical analysis of mistakes (enumerative technique) or by generating mistakes from a set of common misconceptions (generative technique). In enumerative technique, the designers and analysts of the system determine the possible errors that a student can make (Smith 1998). In generative modeling the system uses a cognitive model, which considers students' behavior, to detect students' errors (Clancey 1988).

1.2.3.2 Constraint-Based Model

The Constraint-Based Model (CBM) uses constraints to represent both domain and student knowledge. The knowledge domain is represented as set of constraints and the student model is the set of constraints that have been violated (Fig. 1.6). A constraint has a satisfaction clause and a relevance condition. If the satisfaction clause becomes false for the relevance condition, then the learner has made an error (Martin 1999). The particular model is based on Ohlsson's theory of learning from errors (Ohlsson 1996). According to this theory a learner often makes mistakes when performing a task, even when s/he has been taught the correct way to do it. According to Mitrovic et al. (2001), the most important advantages of CBM are: its computational simplicity, the fact that it does not require a runnable expert module, and the fact that it does not require extensive studies of student bugs as in enumerative modeling.

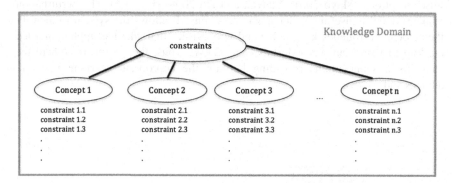

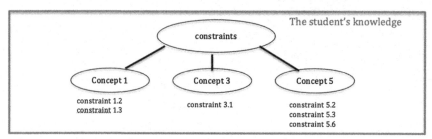

Fig. 1.6 Constraint-based model

1.2.4 Machine Learning Techniques

The student model is responsible for the identification of the student's knowledge level, misconceptions, needs and preferences. This kind of information is obtained by observing the student's behavior and action during her/his interaction with the adaptive and/or personalized tutoring system. The processes of the student's behavior observation and reasoning should be made automated by the system. This is achieved by machine learning techniques. Machine learning concerns the formation and study of models that allow the system to learn from observation's data and make automatically inferences (Webb 1998). Machine learning have so far been used either to induce a single, consistent student model from multiple observed student behaviors, or for the purpose of automatically extending or constructing from scratch the bug library of student modelers (Sison and Shimura 1998). Therefore, machine-learning techniques can be used to predict future actions (Webb et al. 2001) and make the system able to adapt the instruction and learning processes to the student's needs. An approach of machine learning is the use of artificial neural networks. They are computational systems inspired by the biological nervous system of the brain. Artificial neural networks are presented as interconnected networks of "neurons" that can learn through experience via algorithms.

1.2.5 Cognitive Theories

The adaptive and/or personalized tutoring systems have to integrate pedagogical and psychological theories, except of artificial intelligence, to be effective. Indeed, many researchers (e.g. Salomon 1990; Welch and Brownell 2000) have pointed out that technology is effective when developers thoughtfully consider the merit and limitations of a particular application while employing effective pedagogical practices to achieve a specific objective. Pedagogical practices can be integrated in a student model by using cognitive theories, which attempt to explain human behavior during the learning process. Cognitive theories can model either the student's cognitive characteristics like knowledge, attention, ability to learn and understand and memory or the student's emotional states and motivation. Therefore, they contribute significantly to the student's reasoning trying to understand human's processes of thinking and understanding.

There are a variety of cognitive theories. Some cognitive theories that have been used in student modeling are: the Human Plausible Reasoning (HPR) theory (Collins and Michalski 1989), which is a domain-independent theory that categorizes plausible inferences in terms of a set of frequently recurring inference patterns and a set of transformations on those patterns (Burstein and Collins 1988; Burstein et al. 1991); the Ortony et al. (1988) (OCC) theory, which allows modeling possible emotional states of students, and the Control-Value theory (Pekrun et al. 2007), which is an integrative framework that employs diverse factors, e.g. cognitive, motivational and psychological, to determine the existence of achievement emotions. The use of cognitive theories in student modeling adds more "human" reasoning to the computer.

1.2.6 Modeling the Uncertainty of Learning

The processes of learning and student's diagnosis are complex. They are defined by many factors and are depended on tasks and facts that are uncertain and, usually, unmeasured. The determination of the student's knowledge, mental state and behavior is not a straightforward task, but it is based on uncertain observations, measurements, assumptions and inferences. The presence of uncertainty in student's diagnosis is increased in an adaptive/personalized tutoring system due to either the indirect interaction between the learner and the teacher, or the technical difficulties (Grigoriadou et al. 2002). The most common used techniques to encounter this kind of uncertainty are fuzzy logic and Bayesian Networks.

1.2.6.1 Fuzzy Student Modeling

Fuzzy logic was introduced by Zadeh (1965) as a methodology for computing with words. It is able to handle the uncertainty of learning and student's diagnosis, which is based on imprecise data and human decisions, since it encounters the uncertainty problems that are caused by incomplete data and human subjectivity (Drigas et al. 2009). The core of the fuzzy logic theory is the fuzzy sets, which are used to describe an element (characteristic, thing, fact or state) and have no concrete limits (Fig. 1.7). An element can belong to two adjacent fuzzy sets at the same time, but with different membership degree. For example, a student can be 85 % advanced (membership degree: 0.85) and 15 % expert (membership degree: 0.15) or 30 % novice (membership degree: 0.3) and 70 % beginner (membership degree: 0.7) (Fig. 1.8).

Fuzzy logic can help to improve the adaptation of an intelligent tutoring system. Fuzzy logic can help the system to decide what is the appropriate instruction model

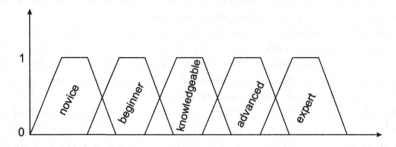

Fig. 1.7 Fuzzy sets of the students knowledge level

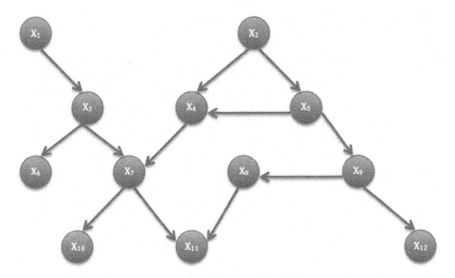

Fig. 1.8 A Bayesian network

each time considering a set of criteria and model specifications (Shakouri and Menhaj 2008). Chrysafiadi and Virvou (2012) have showed that the integration of fuzzy logic into the student model of an ITS can increase learners' satisfaction and performance, improve the system's adaptivity and help the system to make more valid and reliable decisions. Therefore, several researchers have incorporated fuzzy logic techniques in student modeling.

1.2.6.2 Bayesian Networks

Bayesian Networks is another well-established tool for representing and reasoning about uncertainty in student models (Conati et al. 2002). BN's graphical representation, sound mathematical foundations and ability to represent uncertainty using probabilities make them attractive to many researchers (Jameson 1996; Liu 2008; Desmarais and Baker 2012). Indeed, the presence of capable and robust Bayesian libraries (e.g. SMILE), which can be easily integrated into the existing or new student modeling applications, facilitates the adoption of BNs in student modeling (Millán et al. 2010). A Bayesian Network (BN) is a directed acyclic graph in which nodes represent variables and arcs represent probabilistic dependence or causal relationships among variables (Pearl 1988) (Fig. 1.8). The causal information encoded in BN facilitates the analysis of action sequences, observations, consequences and expected utility (Pearl 1996). In student modeling nodes of a BN can represent the different components/dimensions of a student such as knowledge, misconceptions, emotions, learning styles, motivation, goals etc.

1.2.7 Ontology-Based Student Modeling

Recently a lot of research has been done on the crossroad of user modeling and web ontologies. Due to the fact that the adaptive and/or personalized tutoring systems attempt to model the teaching and learning processes in real world and the most of them are web-based applications, they can be combined with web ontologies. Ontologies support the representation of abstract enough concepts and properties and make them reused and extended in different application (Clemente et al. 2011). These characteristics of ontologies can help student modeling. The main advantages of ontology-based student models are: formal semantics, easy reuse, easy probability, availability of effective design tools, and automatic serialization into a format compatible with popular logical inference engines (Winter et al. 2005).

1.3 Student's Characteristics to Model

A significant initial stage of constructing a student model is the selection of appropriate students' characteristics that should be considered and represented. The personalization is accomplished efficiently by modeling either the domain dependent student's characteristics or the domain independent domain student's characteristics (Yang et al. 2010). For example, domain dependent student's characteristics are the knowledge level, the misconceptions, and the prior knowledge. Some domain independent student's characteristics are learning style, memory, concentration, and self-assessment. The student's characteristics are, also, categorized into static characteristics (like email, age, native language) and dynamic characteristics (like knowledge level, errors). The static characteristics are set by the student at the beginning of the learning process, usually through questionnaires, while the dynamic characteristics are defined and updated each time the student interacts with the system.

Therefore, the challenge is to define the dynamic student's characteristics that constitute the base for the system's adaptation to each individual student's needs. These characteristics include knowledge and skills, errors and misconceptions, learning styles and preferences, affective and cognitive factors, meta-cognitive factors (Fig. 1.9).

The student's characteristics that are usually modeled are:

1. Knowledge level
2. Errors and misconceptions
3. Cognitive features other than knowledge level
4. Affective features
5. Meta-cognitive features

These students' characteristics are described in detail in the following subsections.

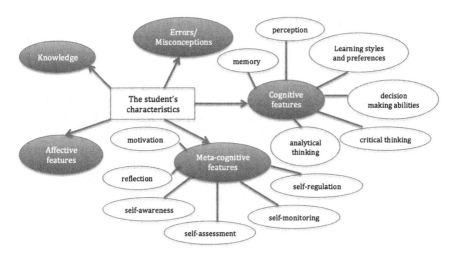

Fig. 1.9 Dynamic characteristics of a student

1.3.1 Knowledge Level

Knowledge level is the most commonly modeled student's characteristic. Knowledge refers to the prior knowledge of a student on the knowledge domain as well as her/his current knowledge level. The student's knowledge level is usually assessed through questionnaires and tests that the student has to complete during the learning process.

The most commonly used approach for representing the student's knowledge level is the overlay student model. During last years many adaptive and/or personalized tutoring systems have used overlay to represent the student's knowledge level. Surjono and Maltby (2003) have modeled the learner's knowledge using the overlay approach. Similarly, Kassim et al. (2004) used an overlay student model in a web-based intelligent learning environment for digital systems (WILEDS) in order to represent dynamically the emerging knowledge and skills of each student. Furthermore, in MEDEA (Carmona and Conejo 2004) an estimation of the student knowledge level for each domain concept is stored in an overlay model. InfoMap has used an overlay technique to model the knowledge level of children to basic arithmetic (Lu et al. 2005, 2007). Similarly, an overlay approach has been used in TANGOW (Alfonseca et al. 2006) for modeling the student's knowledge. Also, Kumar (2006a, b) has used an overlay model to represent the knowledge of students in programming tutors. Glushkova (2008) has applied a qualitative overlay student model to represent learners' knowledge level to DeLC system. LS-Plan (Limongelli et al. 2009) is another adaptive learning system that has used a qualitative overlay student model. An overlay model has been used, also, by Mahnane et al. (2012). In addition, PDinamet (Gaudioso et al. 2012), which is a web-based adaptive learning system for the teaching of physics in secondary education, have incorporated an overlay student model to represent the student's knowledge level in order to provide effective and personalized selection of the appropriate learning resources.

Another student modeling technique that is usually used to model the learner's knowledge level is stereotyping. Examples of adaptive and/or personalized tutoring systems that have used stereotypes for modeling the student's knowledge lever are the following. AUTO-COLLEAGUE (Tourtoglou and Virvou 2008, 2012), which is an adaptive and collaborative learning environment for UML, represents the level of students' expertise through a stereotype-based modeling technique. Furthermore, Chrysafiadi and Virvou (2008) have developed a stereotyping approach to model the knowledge level of learners in the programming language Pascal in order to adapt the system's responses to each individual student dynamically. Also, a stereotype-like approach for modeling the student's knowledge level is used in Wayang Outpost, which is a software tutor that helps students learn to solve standardized-test type of questions, in particular for a math test called Scholastic Aptitude Test, and other state-based exams taken at the end of high school in the USA, in order to discern factors that affect student behavior beyond cognition (Arroyo et al. 2010). Moreover, Durrani and Durrani (2010) have used stereotypes for modeling the student's knowledge the adaptive C++ tutor CLT. Finally, Grubišić et al. (2013) have defined knowledge stereotypes based to model the student's proficiency in an adaptive e-learning system called Adaptive Courseware Tutor (AC-ware Tutor).

Another technique of modeling the learner's knowledge is the Constraint-Based Model (CBM). Mitrovic (2003) have used the CBM approach to model the student's knowledge of a web-enabled intelligent tutoring system that teaches the SQL database language. Another system that uses CMB for modeling the student's knowledge is COLLECT-UML, which is an ITS that teaches object-oriented design using Unified Modeling Language (Baghaei et al. 2005). Also, Weerasinghe and Mitrovic (2011) have applied CBM to model the student's knowledge in EER-Tutor, which is an ITS that teaches conceptual database design.

Furthermore, BNs have been used for the representation of the student's knowledge. For example, Bunt and Conati (2003) used Bayesian Networks to detect when the learner is having difficulties in an intelligent exploratory learning environment for the domain of mathematic functions. A Bayesian student model was applied in English ABLE for modeling the student's knowledge in English grammar (Zapata-Rivera 2007). Furthermore, in TELEOS a Bayesian network based student model was used in order to explicitly diagnose the student's knowledge level (Chieu et al. 2010). Similarly, AdaptErrEx has used BNs to model learners' skills (Goguadze et al. 2011a, b). Also, INQPRO system predicts the acquisition of scientific inquiry skills by modeling students' characteristics with Bayesian networks (Ting and Phon-Amnuaisuk 2012).

Several student models for learners' knowledge representation have been built based on ontologies. For example, MAEVIF (Clemente et al. 2011) and SoNITS (Nguyen et al. 2011) have used ontologies to model the student's knowledge. Also, Peña and Sossa (2010) have adopted a semantic representation and management of student models with ontologies in order to represent learners' knowledge.

Moreover, Pramitasari et al. (2009) have developed a student model ontology based on student performance.

However, many adaptive and/or personalized tutoring systems perform modeling of the student's knowledge by combining different student modeling techniques. Thereby, there are systems, like TADV (Kosba et al. 2003, 2005), which combine an overlay model with fuzzy techniques to represent the knowledge of individual students. Stathacopoulou et al. (2005) have used fuzzy techniques to represent the knowledge and abilities of students to help them to construct the concepts of vectors in physics and mathematics. Another combination of the overlay technique is with stereotypes. ICICLE (Michaud and McCoy 2004) is an adaptive tutoring system that attempts to capture the user's mastery of various grammatical units and to predict the grammar rules s/he is most likely using when producing language by combining overlay with stereotypes. A similar combination of student modeling techniques has been performed in ELaC (Chrysafiadi and Virvou 2013c), which is a web-based educational system that teaches the programming language 'C'.

Also, there are adaptive educational systems that have combines overlay and stereotypes with fuzzy techniques to model the learner's knowledge. Examples of such systems are: INSPIRE (Grigoriadou et al. 2002; Papanikolaou et al. 2003), which is an intelligent system for personalized instruction in a remote environment, that models knowledge on a topic classifying it to one of the four levels of proficiency (insufficient, rather insufficient, rather sufficient, sufficient); DEPTHS (Jeremić et al. 2012), which is an intelligent tutoring system for learning software design patterns, models the student's mastery; and FuzKSD (Chrysafiadi and Virvou 2014) that is an e-learning environment for the computer programming.

Other adaptive educational systems that used hybrid student models for representing the student's knowledge level are: KERMIT (Suraweera and Mitrovic 2004), which maintains a constraint-based model and an overlay model; InterMediActor (Kavčič 2004a) that models the student's knowledge using overlay in combination with ontologies; F-SMILE (Virvou and Kabassi 2002) that uses a novel combination of the cognitive theory Human Plausible Reasoning (Collins and Michalski 1989) and a stereotype-based mechanism; and AMPLIA (Viccari et al. 2009) that models the learner's knowledge by combining Bayesian networks with cognitive theories. Furthermore, OPAL (Cheung et al. 2010) and IWT (Albano 2011) used a combination of overlay and ontologies to model the learner's knowledge level.

In addition, a variety of adaptive tutoring systems, like SimStudent (Li et al. 2011) and AIWBES (Homsi et al. 2008), used machine-learning techniques to observe the student's behavior and make inferences about her/his knowledge automatically. Baker et al. (2010) have used a combination of machine learning technique with Bayesian networks in order to observe students' reactions and adjust the instruction automatically to each individual learner. Furthermore, Web-EasyMath (Tsiriga and Virvou 2002, 2003c), which is a Web-based Algebra Tutor, uses a combination of stereotypes with the machine learning technique of the distance weighted k-nearest neighbor algorithm, in order to initialize the model of a

Table 1.1   Student modeling approaches in relation to knowledge level

	Overlay	Stereotypes	Constraint-based model	Machine learning	Cognitive theories	Fuzzy techniques	Bayesian networks	Ontologies
Knowledge level	42.55 %	29.79 %	8.51 %	14.89 %	4.26 %	10.64 %	14.89 %	14.89 %

new student. The student is first assigned to a stereotype category concerning her/his knowledge level and then the system initializes all aspects of the student model using the distance weighted k-nearest neighbor algorithm among the students that belong to the same stereotype category with the new student. A combination of stereotypes with machine learning techniques has been, also, used in Web-PTV (Tsiriga and Virvou 2003a, b) and GIAS (Castillo et al. 2009) to model the learner's knowledge. Moreover, Al-Hmouz et al. (2010, 2011) have applied a hybrid student model, which combines machine-learning techniques with stereotypes, to predict the student knowledge.

Therefore, there are a variety of student modeling techniques that can be used or combined to model the learner's knowledge. Each one is preferred in relation with the system's characteristics and the researchers needs. In Table 1.1, the percentages of preferences for each one of the student modeling techniques for modeling the student's knowledge are presented considering the above literature review. The information that is derived from the particular table is the number of the adaptive educational systems that incorporate a particular student modeling technique for the representation of the learner's knowledge level in a set of one hundred adaptive educational systems. For example, if we have a hundred adaptive educational systems 42.55 of them will use overlay, 29.79 will use stereotypes etc. A system can integrate more than one student modeling techniques.

1.3.2 Errors/Misconceptions

Knowledge level is not the only the common student's characteristic that is, usually, detected and measured through questionnaires and tests. The educational system can, also, identify the student's misconceptions and errors through these tests as well as observing student's actions during the learning process. A student's misconception is an erroneous belief, idea, thought. It is a misunderstanding that is usually caused by incorrect thinking or faulty facts.

Many researchers of intelligent and/or adaptive educational systems have modeled the student's errors and misconceptions in order to provide each individual learner with personalized feedback and support. The most commonly used approach for modeling the learner's errors and misconceptions is the perturbation model. Many adaptive educational systems have used the particular student modeling technique perturbation to model the student's errors and misconceptions. In particular, Surjono and Maltby (2003) have used a perturbation student model to perform a better remediation of student mistakes. Furthermore, LeCo-EAD (Faraco et al. 2004) and InfoMap (Lu et al. 2005) have modeled students' misconceptions by using a perturbation model. InfoMap's perturbation student model involves 31 types of addition errors and 51 types of subtraction errors (Lu et al. 2005). The student model of both systems allows the reasoning of students' errors and helps the system to expand the explanation during the feedback to the students. Moreover, Baschera and Gross (2010) have represented through the perturbation approach the student's strength and weaknesses, in order to allow for appropriate remediation actions to adapt to students' needs. A perturbation student model for detecting the student's errors has been used in AUTO-COLLEGE (Tourtoglou and Virvou 2012).

Furthermore, there are adaptive tutoring systems that have used other techniques than perturbation to model the student's errors and misconceptions. In particular, Virvou and Kabassi (2002) have added more "human" reasoning to F-SMILE by using stereotypes and cognitive theory of Human Plausible Reasoning (HPR) (Collins and Michalski 1989). F-SMILE reacts accordingly trying to find out the cause of the problematic situation in which the user is involved when s/he learns how to manipulate file store of her/his computer. Goel et al. (2012) used a fuzzy model for student reasoning based on imprecise information coming from the student-computer interaction and performed the prediction of the degree of error a student makes in the next attempt to a problem. Also, Chrysafiadi and Virvou (2008) have modeled the type of programming errors that a student can make during her/his interaction with a web-based educational application that teaches the programming language Pascal (Web_Tutor_Pas) using stereotypes. Furthermore, so KERMIT (Suraweera and Mitrovic 2004) that teaches conceptual database design, as J-LATTE (Holland et al. 2009) and INCOM (Le and Menzel 2009), which teach language programming, use the CBM approach to diagnose the student's errors. In addition, AdaptErrEx has used BNs to model learners' misconceptions (Goguadze et al. 2011a, b). BNs have been, also, used for modeling student's errors in Andes (Shapiro 2005). Moreover, Pérez-de-la-cruz (2002) has modeled the student's misconceptions applying BNs in combination with cognitive theories.

Therefore, there are a variety of student modeling techniques that can be used to model the learner's errors and misconceptions. In Table 1.2, the percentages of preferences for each one of the student modeling techniques for modeling the student's errors and misconceptions are presented considering the above literature review. The information that is derived from the particular table is the number of the adaptive educational systems that incorporate a particular student modeling technique for the modeling of the learner's errors and misconceptions in a set of one hundred adaptive educational systems. For example, if we have a hundred adaptive educational systems 35.71 of them will use the perturbation approach, 21.43 will use the Constraint-based model, 21.43 % will use Bayesian Networks etc.

Table 1.2   Student modeling approaches in relation to student's errors and misconceptions

	Perturbation	Constraint-based model	Stereotypes	Cognitive theories	Fuzzy techniques	Bayesian networks
Errors/ misconceptions	35.71 %	21.43 %	14.29 %	14.29 %	7.14 %	21.43 %

1.3.3 Cognitive Features Other Than Knowledge Level

Student's cognitive features are among the most sophisticated student character-istics that are described in a student model. These features refer to aspects such as attention, ability to learn and understand, memory, perception, concentration, collaborative skills, abilities to solve problems and making decisions, analyzing abilities, critical thinking, learning style and preferences.

Learning style refers to individual skills and preferences that affect how a stu-dent perceives, gathers and processes learning materials (Jonassen and Grabowski 1993). Some learners prefer graphical representations, others prefer audio materials and others prefer text representation of the learning material, some students prefer to work in groups and others learn better alone Popescu (2009). Adapting courses to the learning preferences of the students has a positive effect on the learning pro-cess, leading to an increased efficiency, effectiveness and/or learner satisfaction (Popescu et al 2010). A proposal for modeling learning styles, which are adopted by many ITSs, is the Felder-Silverman learning style (FSLSM). FSLSM classifies students in four dimensions: active/reflective, sensing/intuitive, visual/verbal, and sequential/global (Felder and Silverman 1988; Felder and Soloman 2003). Another method for modeling learning styles is the Myers-Briggs Type Indicator (MBTI) (Bishop and Wheeler 1994), which identifies the following eight categories of learning styles: extrovert, introvert, sensing, intuitive, thinking, feeling, judging, perceiving.

Many researchers have modeled the student's learning style and preferences. Most of them have been used stereotypes for modeling the particular cogni-tive features. For example, the stereotypes of the student model of INSPIRE (Grigoriadou et al. 2002; Papanikolaou et al. 2003) provides information about the learning style of the learner. Furthermore, Surjono and Maltby (2003) have used stereotypes to model the student's preferences (i.e. font, colour, illustration) and learning styles (i.e. competitive, collaborative, avoidant, participant, depend-ent, independent). Also, Glushkova (2008) has modeled the student's preferences, habits and behaviors during the learning process by using stereotypes. Moreover, Carmona et al. (2008) have used a student model that classifies students in four stereotypes according to their learning styles. In WELSA (Popescu et al. 2009) the courses are adapted to the learning preferences of each student applying stereotyping.

In addition, Salim and Haron (2006) used a combination of fuzzy logic with a stereotype-like mechanism to model the student's personality factor MBTI (Myers-Briggs Type Indicator). AHA! (Stash et al. 2006) and TANGOW

(Alfonseca et al. 2006) have modeled the student's learning styles using an overlay approach. Fuzzy techniques have been, also, used by Stathacopoulou et al. (2005) for modeling the student's learning style. They have applied a student model to a discovery-learning environment that aimed to help students to construct the concepts of vectors in physics and mathematics, which drive pedagogical decisions depending on the student learning style. Furthermore, Crockett et al. (2013) have tried to predict learning styles in a conversational intelligent tutoring system using fuzzy logic. Similarly, Oscal CITS adapts to the student's learning styles incorporated a fuzzy mechanism (Latham et al. 2014). Also, TADV (Kosba et al. 2003, 2005) includes a student model, which combines overlay with fuzzy logic, to represent communication styles of individual students, except of their knowledge.

Moreover, in GIAS (Castillo et al. 2009) the appropriate selection of the course's topics and learning resources are based not only on the student's goals and knowledge level but also on the student's learning style that is modeled using stereotypes and machine learning techniques. In addition, many researchers, like Bunt and Conati (2003), Parvez and Blank (2008), Schiaffino et al. (2008), and Bachari et al. (2011) have been used Bayesian Networks to detect a student's learning style and/or preferences automatically. To perform the same goal, Hernández et al. (2010) have combined Bayesian Networks with cognitive theories. In addition, Lo et al. (2012) as well as Zatarain-Cabada et al. (2010) have used artificial neural networks (learning machine) to identify the student's cognitive and learning styles correspondingly. Finally, the student's preferences in Personal Reader (Dolog et al. 2004) and in the tutoring system of Pramitasari et al. (2009) have been modeled by using ontologies.

Many attempts to model other cognitive characteristics of students except of learning styles have, also, made. Conati et al. (2002) have tried to model in Andes cognitive aspects like long-term knowledge assessment, plan recognition, ability to solve problems and reading latency using Bayesian Networks. In Web-PTV (Tsiriga and Virvou 2003a, b), which teaches the domain of the passive voice of the English language, the carefulness of the student while solving exercises is estimated through a hybrid student model, which combines stereotypes with machine learning techniques. Furthermore, in F-CBR-DHTS (Tsaganoua et al. 2003) the diagnosis of students' cognitive profiles of historical text comprehension was done with fuzzy techniques and a stereotype-like mechanism. In TELEOS the student's cognitive behavior has been explicitly diagnosed through Bayesian Networks (Chieu et al. 2010). AUTO-COLLEAGUE (Tourtoglou and Virvou 2008, 2012) uses stereotypes to model the personality of students. Durrani and Durrani (2010) have considered the student's cognitive abilities in the adaptive C++ tutor CLT using stereotypes, also. Jia et al. (2010) have designed an adaptive learning system, which is based on fuzzy logic and helps learners to memory the content and improve their comprehension. Peña and Sossa (2010) have used an ontology-based student model to represent learners' knowledge, personality, learning preferences and content, and to deliver the appropriate option of lecture to students.

Furthermore, DEPTHS (Jeremić et al. 2012), which is an intelligent tutoring system for learning software design patterns, models, except of the student's

Table 1.3 Student modeling approaches in relation to student's learning style and preferences

	Overlay	Stereotypes	Fuzzy techniques	Cognitive theories	Machine learning techniques	Bayesian networks	Ontologies
Learning styles and preferences	13.64 %	31.82 %	22.73 %	4.55 %	13.64 %	22.73 %	4.55 %

mastery, her/his cognitive characteristics by combining overlay with stereotypes and fuzzy techniques. Also, Mahnane et al. (2012) have used stereotypes to integrate thinking style (AHS-TS) in an adaptive hypermedia system. In addition, Wang et al. (2009) have built a student model, which is based on machine learning techniques and represents the learner's language competence, cognitive characteristics and learning preferences, in order to assist students in successfully mastering the English language. Other researchers that have modeled the cognitive characteristics of students are: Jurado et al. (2008), who have used machine learning techniques in combination with fuzzy techniques; Al-Hmouz et al. (2010, 2011), who have combined stereotypes with machine learning techniques, and Viccari et al. (2008), who have built a student model based on cognitive theories and Bayesian networks.

Therefore, there are a variety of student modeling techniques that can be used to model the learner's cognitive features. In Table 1.3, the percentages of preferences for each one of the student modeling techniques for modeling the student's learning styles and preferences are presented considering the above literature review. Furthermore, in Table 1.4, the percentages of preferences for each one of the student modeling techniques for modeling the student's general cognitive features other than knowledge (including learning styles and preferences) are presented considering the above literature review. The information that is derived from the above tables is the number of the adaptive educational systems that incorporate a particular student modeling technique for modeling the student's learning style, preferences and other cognitive features in a set of one hundred adaptive educational systems. From the data on the tables, it is concluded that stereotypes is the most popular student modeling technique for representing the student's learning styles and other cognitive features (other than knowledge).

Table 1.4 Student modeling approaches in relation to student's cognitive features other than knowledge

	Overlay	Stereotypes	Fuzzy techniques	Cognitive theories	Machine learning techniques	Bayesian networks	Ontologies
Cognitive features other than knowledge	8.33 %	38.89 %	22.22 %	5.56 %	19.44 %	22.22 %	8.33 %

1.3.4 Affective Features

The emotional state of a student affects the learning process and the student's performance and progress. The emotional state can have a negative or positive effect on learning. That is the reason why in real classroom settings, experienced teachers and professors observe and react accordingly to the emotional state of the students in order to motivate them and improve their learning process (Johnson et al. 2000; Lehman et al. 2008). Therefore, adaptive and/or personalized educational systems should detect the emotional state of students and adapt its behavior to their needs, giving an appropriate response for those emotions (Katsionis and Virvou 2004).

These emotional factors that influence learning are called affective factors. The affective states can be the following: happiness, sadness, anger, anxiety, interest, fear, boredom, frustration, distraction, confusion, tiredness, indifference, concentration and enthusiasm. Some of these emotions, like happiness and concentration, have positive effect on the learning process. However, other emotions, like boredom, tiredness and distraction, have negative effect on the learning process and lead students to an off-task behavior (Rodrigo et al. 2007), which are associated, usually, with deep motivational problems (Baker 2007). Off-task behavior means that students' attention becomes lost and they engage in activities that have anything to do with the learning process and aim (Cetintas et al. 2010), like surfing the web, devoting time to off-topic readings, talking with order students without any learning aims (Baker et al. 2004). Therefore, the affective factors should be considered when a student model is built.

Many researchers have used cognitive, pedagogical and psychological theories in combination with student modeling techniques in order to identify and model the emotional states of students. In particular, Conati and Zhou (2002) have used the OCC cognitive theory of emotions (Ortony et al. 1988) for recognizing user emotions for their educational game prime climb. The same theory has also been used in a Mobile Medical Tutor (MMT) for modeling possible states that a tutoring agent may use for educational purposes (Alepis and Virvou 2011). The same researchers have constructed user stereotypes concerning the users emotional behavior while they interact with computers (Alepis and Virvou 2006). VIRGE is another ITS-game, which has adopted OCC theory in order to provide important evidence about students' emotions while they learn (Katsionis and Virvou 2004; Virvou et al. 2005).

A significant attempt to recognize and convey emotions in order to enhance students' learning and engagement have been done by Muñoz et al. (2010, 2011) in PlayPhysics, which is an emotional game-based learning environment for teaching physics. They have used Bayesian networks in combination with the Control-Value theory (Pekrun et al. 2007), which is an integrative framework that employs diverse factors, e.g. cognitive, motivational and psychological, to determine the existence of achievement emotions. Furthermore, Alepis et al. (2008) have described a novel mobile educational system that incorporates bimodal emotion

Table 1.5 Student modeling approaches in relation to student's affective features

	Stereotypes	Machine learning techniques	Cognitive theories	Bayesian networks	Ontologies
Affective features	6.67 %	40 %	40 %	33.33 %	6.67 %

recognition through a multi-criteria theory. Also, Conati and Mclaren (2009) developed a probabilistic model of user affect, which recognizes a variety of user emotions by combining information on both the causes and effects of emotional reactions. Moreover, Moridis and Economides (2009) have developed a neural network method (machine learning technique) to recognize a learner's affective state.

Also, Baker (2007) have constructed a machine learning based model that can automatically detect when a student using an intelligent tutoring system is off-task, i.e. engaged in behavior, which does not involve the system or a learning task. Similarly, Cetintas et al. (2010) have performed the automatic detection of off-task behaviors in intelligent tutoring systems using machine-learning techniques. Furthermore, Balakrishnan (2011) build a student model upon ontology of machine learning strategies in order to model the effect of affect on learning. Machine learning techniques have been also used for predicting the emotions of boredom and curiosity in an Intelligent Tutoring System that is called MetaTutor (Jaques et al. 2014). Also, Hernández et al. (2010) have applied an affective student model combining the OCC theory with Bayesian Networks. Inventado et al. (2010) have used a combination of Bayesian Networks and machine learning techniques to model the student's affective features in POOLE III. Finally, Crystal Island, which is a game-based learning environment, uses Bayesian Networks to model and predict student affect for improving the learning process and motivation.

Therefore, there are a variety of student modeling techniques that can be used to model the learner's affective features. In Table 1.5, the percentages of preferences for each one of the student modeling techniques for modeling the student's affective features are presented considering the above literature review. The information that is derived from the particular table is the number of the adaptive educational systems that incorporate a particular student modeling technique for modeling the learner's emotions in a set of one hundred adaptive educational systems. For example, if we have a hundred adaptive educational systems 40 of them will use cognitive theories and machine learning techniques, 33.33 will use Bayesian networks etc.

1.3.5 Meta-Cognitive Features

Meta-cognitive features allow the student to be aware of her/his knowledge and abilities and make her/him able to monitor and direct her/his own learning processes. In other words, meta-cognition concerns to the active monitoring, controlling, regulation and orchestration (Flavell 1976). For example, a student has meta-cognitive features when s/he is aware of and controls their own thinking; s/he is able to select her/his own learning goal; s/he can use properly the obtained and prior knowledge; s/he can choose the appropriate each time problem-solving strategy (Mitrovic and Martin 2006; Barak 2010). Some meta-cognitive skills are reflection, self-awareness, self-monitoring, self-regulation, self-explanation, self-assessment, and self-management (Peña and Kayashima 2011). Metacognitive features allow the students to participate more actively to her/his own learning processes. Thereby, adaptive and/or personalized tutoring systems must consider students' meta-cognitive skills.

Although the field of meta-cognitive students' features is a new field of research, a number of attempts to model the meta-cognitive feature of students have been made. Conati et al. (2002) have tried to monitor and encourage self-explanation of students that learn Newtonian physics using Bayesian networks. Wayang Outpost, which is a software tutor that helps students learn to solve standardized-test type of questions, have tried to recognize students' behaviors related to meta-cognitive factors (Arroyo et al. 2010). Furthermore, Albano (2011) has presented a model that allows the students to build up their competence in mathematics concerning meta-cognitive factors. Moreover, Ting et al. (2013) have modeled student engagement in a computer-based scientific inquiry learning environment using a Bayesian Networks model. Also, Liaw and Huang (2013) and Cho and Kin (2013) have investigated learner self-regulation in e-learning environment. They have tried to define the learner's characteristics that affect the self-regulation.

Meta-cognitive features are very sophisticated and constitute a new field of research. Their identification is difficult and their definition is complex. There is not adequate literature review to be able to draw conclusions about either the learner's characteristics that determine a meta-cognitive factor or the technique that is suitable for modeling a particular meta-cognitive feature. That is the reason for the absence of a comparative discussion about the meta-cognitive features in relation with student modeling techniques from this subsection.

1.4 Discussion

Student modeling is a research field of the Intelligent Tutoring Systems (ITSs) that has attracted the interest of many researchers. Although, student modeling has been introduced in ITSs, its use has been extended to most current educational software applications that aim to be adaptive and personalized. Therefore, an attempt to model the student's characteristics has been made in many adaptive educational systems. The aim of each adaptive educational system is to model the most appropriate student's characteristics in order to carry out the personalization efficiently. Hence, the student's characteristics that are usually modeled are: the student's knowledge and misconceptions, her/his preferences and cognitive

Table 1.6 Combination for compound student model

	Overlay	Stereotypes	Erroneous knowledge models	Machine learning	Cognitive theories	Uncertainty models	Ontology-based models
Overlay		x	x			x	x
Stereotypes	x		x	x	x	x	
Erroneous knowledge models	x	x					
Machine learning		x			x	x	x
Cognitive theories		x		x		x	
Uncertainty models	x	x		x			
Ontology-based models	x			x	x		

features, affective and meta-cognitive factors. The developers of the student model select the most appropriate each time student-modeling technique or the most appropriate combination of such techniques to model the above student's characteristics. The prevailing student modeling techniques that are presented in the literature are: overlay, stereotypes, perturbation model, constraint-based model, machine learning techniques, neural networks, cognitive theories, fuzzy logic techniques, Bayesian networks and ontologies. Many adaptive and/or personalized tutoring systems perform student modeling combing different modeling techniques to bring together features of different techniques of user modeling. A compound student model allows the tutoring system to carry out the personalization efficiently. Table 1.6 presents the most common combination of student modeling techniques. In that table the category of erroneous knowledge models include the perturbation and the Constraint-based model. Also, the category of uncertainty models includes fuzzy logic techniques and Bayesian networks.

Two questions come of the literature review: (i) "Which student's characteristics are the most common-modeled?" (ii) "Which student modeling approaches are preferred in relation to student modeling characteristics?". A thorough study and comparison of the adaptive and/or personalized tutoring systems that were presented in this chapter give answers to the above two questions. The presented adaptive and/or personalized tutoring systems have been developed from 2002 up to now (2014). Mostly of them (96 %) are results of Scopus, which is the world's largest abstract and citation database of peer-reviewed literature. Scopus is considered as one of the most valid search engine for research papers. Furthermore, a respectable number of these systems have been evaluated. The rest systems, which have not been evaluated, are trends that have not been established yet.

The results of this literature review are very interesting and useful for the researchers, designers and developers of educational systems and student models.

Table 1.7 Which student's characteristics are preferred for modeling

Knowledge level	Error/misconceptions	Cognitive features other than knowledge
52.81 %	15.73 %	40.45 %

Affective features		Meta-cognitive features
16.85 %		6.74 %

According to that, the most common-modeled student's characteristic is the knowledge level and the least common-modeled student's characteristic is her/his meta-cognitive features (Table 1.7). The sum of the percentages of Table 1.7 is not 100 %. The reason for this is the fact that a system can model more than one different student characteristics. Also, many researchers have interested in modeling student's cognitive aspects other than knowledge. Furthermore, the answer to the question "Which student modeling approaches are preferred in relation to student modeling characteristics?" is given in Table 1.8. The sum of the percentages of a line of Table 1.8 is not 100 %. The reason for this is the fact that two different student-modeling techniques can be combined and used in the same system. For example, a system can combine stereotypes with machine learning techniques to model the student's learning style. The information that is derived from each line of the particular table is the answer to the question: "if there are one hundred adaptive educational systems how many of them will incorporate a particular student modeling technique to model the learner's characteristic that corresponds to the table's line?". The results of the research (Table 1.8) demonstrated that: (i) the most common used student modeling technique for the representation of the student's knowledge level is the overlay approach; (ii) the perturbation and constraint-based model (erroneous knowledge models) are preferred for representing the student's misconceptions and errors; (iii) uncertainty models (like fuzzy logic techniques and Bayesian networks) and stereotypes are preferred for modeling student's cognitive aspects other than knowledge; (iv) the uncertainty models are, also, chosen to represent the affective and meta-cognitive features of the student; (v) the student's emotions and affective features are very

Table 1.8 Student modeling approaches in relation to student modeling characteristics

	Overlay (%)	Stereotypes (%)	Erroneous knowledge models (%)	Machine learning (%)	Cognitive theories (%)	Uncertainty models (%)	Ontology-based models (%)
Knowledge level	42.55	29.79	8.51	14.89	4.26	25.53	14.89
Errors/misconceptions	0	14.29	57.14	0	14.29	28.57	0
Cognitive features other than knowledge	8.33	38.89	0	19.44	5.56	44.44	8.33
Affective features	0	6.67	0	40	40	33.33	6.67

often modeled with machine-learning techniques, also. Meta-cognitive features of students are not presented in Table 1.8, due to the fact that student modeling of these features is a new field of research and there are not adequate related references in literature. Moreover, the ontology-based student model has not been developed enough, since the research interest in the particular student modeling approach has started to arise recently.

Chapter 2
Fuzzy Logic in Student Modeling

Abstract The significant development of the e-learning systems has changed the ways of teaching and learning. In nowadays, everyone can have access to e-learning systems from everywhere. Therefore, the e-learning systems have to adapt the learning material and processes to the needs of each individual learner. However, learning and student's diagnosis are complex processes, which deal with uncertainty. A solution to this is the use of fuzzy logic, which is able to deal with uncertainty and inaccurate data. This chapter explains how fuzzy logic can be used to automatically model the learning or forgetting process of a student, offering adaptation and increasing the learning effectiveness in Intelligent Tutoring Systems. In particular, it presents a novel rule-based fuzzy logic system, which models the cognitive state transitions of learners, such as forgetting, learning or assimilating. The operation of the presented approach is based on a Fuzzy Network of Related-Concepts (FNR-C), which is a combination of a network of concepts and fuzzy logic. It is used to represent so the organization and structure of the learning material as the knowledge dependencies that exist between the domain concepts of the learning material.

2.1 Introduction

Over the past decade, the rapid development of computer and Internet technologies has affect a variety of fields of the human's everyday life. Such a field is the education. The ways of teaching and learning have been changed and the e-learning systems and processes have been developed significantly. E-learning systems offer easy access to knowledge domains and learning processes from everywhere for everybody at any time. As a result, users of web-based educational systems are of varying backgrounds, abilities and needs. Therefore, the e-learning systems and applications have to offer dynamic adaptation to each individual student.

Adaptation is performed through the student model. In particular, the student model is a core component in any intelligent or adaptive tutoring system that is responsible for identifying and reasoning the student's knowledge level, misconceptions, abilities, preferences and needs. The student model represents many

© Springer International Publishing Switzerland 2015   25
K. Chrysafiadi and M. Virvou, *Advances in Personalized Web-Based Education*,
Intelligent Systems Reference Library 78, DOI 10.1007/978-3-319-12895-5_2

of the student's features, such as knowledge and individual traits, so as to be accessible for offering adaptation (Brusilovsky and Millán 2007). The adaptive and/or personalized educational system consults the student model and delivers the learning material to each individual learner with respect to her/his personal characteristics.

However, student modeling in many cases deals with uncertainty. Learning and student's diagnosis are complex. They are defined by many factors and are depended on tasks and facts that are uncertain and, usually, unmeasured. One possible approach to deal with this is fuzzy logic, which was introduced by Zaheh (1965) as a methodology for computing with words in order to handle uncertainty. It encounters the uncertainty problems that are caused by incomplete data and human subjectivity (Drigas et al. 2009). Chrysafiadi and Virvou (2012) have showed that the integration of fuzzy logic into the student model of an ITS can increase learners' satisfaction and performance, improve the system's adaptivity and help the system to make more valid and reliable decisions. Consequently, fuzzy logic techniques are able to analyze the students' knowledge level, needs and behavior and to make the right decision about the instructional model that has to be applied for each individual learner.

The issue of fuzzy logic and how it can be used in student modeling are presented in the remainder of this chapter. In particular, an overview of the fuzzy logic theory and fuzzy sets are described. Also, applications of fuzzy logic in student modeling are presented. Furthermore, the use of fuzzy logic in the representation of the knowledge domain of an adaptive and/or personalized tutoring system is described. In addition, a novel rule-based fuzzy logic system for modeling automatically the learning or forgetting process of a student is presented. Finally, a brief discussion and the conclusions drawn from this work are presented.

2.2 An Overview of Fuzzy Logic

Fuzzy logic was introduced by Zadeh (1965) to encounter imprecision and uncertainty. It deals with reasoning that is approximate rather than fixed and exact. It is a precise logic of imprecision and approximate reasoning (Zadeh 1975, 1979). In other words, fuzzy logic is able to reason and make rational decisions in circumstances of imprecision, uncertainty, human subjectivity, incomplete information and deficient computations (Zadeh 2001).

The basic element of the fuzzy logic theory is the fuzzy set. A fuzzy set describes a characteristic, thing, fact or state. For example, 'novice' is a fuzzy set that describes the student's knowledge level, 'young' is a fuzzy set that describes the person's age, 'cold' is a fuzzy set that describes the environment's temperature, 'tall' is a fuzzy set that describes the person's height, 'loud' is a fuzzy set that describes the sound's intensity, 'close' is a fuzzy set that describes the distance between two objects. The fuzzy sets that describe an element have no concrete limits (Fig. 2.1).

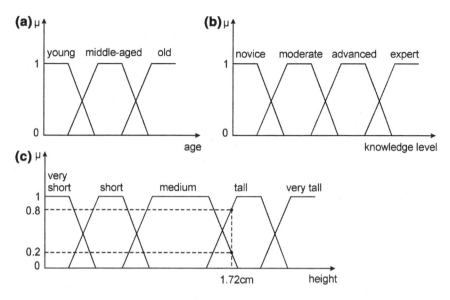

Fig. 2.1 Fuzzy sets and their partitions. **a** Fuzzy sets for age; **b** Fuzzy sets for knowledge level; **c** Fuzzy sets for height

Fuzzy logic variables have a truth-value that ranges in degree between 0 and 1. That value declares the degree in which the particular variable belongs to a fuzzy set. For example, if x is a fuzzy logic variable that describes the student's knowledge level and its value is 0.6 for the fuzzy set 'novice', then it means that the particular student is considered to be 60 % novice. This value is called degree of membership or membership value and is symbolized with μ. A fuzzy logic element can belong to two adjacent fuzzy sets at the same time, but with different membership degrees. For example, if a person's height is 1.72 cm, then according to the fuzzy sets that are depicted in Fig. 2.1c, the particular person is considered to be 80 % tall (the membership degree for the fuzzy set 'tall' is 0.8) and 20 % medium (the membership degree for the fuzzy set 'medium' is 0.2).

Taking into account the above, the definition of a fuzzy set follows (Fig. 2.2). Let S be a set of values that represent an element (i.e. S = {1.20, ..., 2.10} for

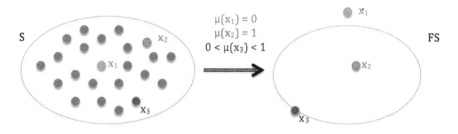

Fig. 2.2 Definition of fuzzy set

height; S = {1, 2, 3, ..., 120} for age; S = {0, 1, 2, ..., 100} for grades) and x ∈ S. In other words, x is a particular value that belongs to the set S. A fuzzy set FS is a pair (x, μ(x)), where x ∈ S and μ(x): S → [0, 1]. In other words, for each x ∈ S, there is a value μ(x) between 0 and 1, which declares the membership degree of x to the fuzzy set FS.

- If μ(x) = 0, then x is not included in FS
- If μ(x) = 1, then x is fully included in FS
- If 0 < μ(x) < 1, then x is partially included in FS

2.2.1 Type-1 Fuzzy Sets

This first approach of fuzzy sets theory, which points that the value of the membership function of a fuzzy set can range between 0 and 1, is called type-1 fuzzy sets. Two common examples of a membership function of type-1 fuzzy sets are depicted in Fig. 2.3. Type-1 fuzzy sets have been criticized about their ability to handle uncertainty. It has been advocated that it is not reasonable to use an accurate membership function for something uncertain. Type-1 fuzzy sets used in conventional fuzzy systems cannot fully handle the uncertainties that are present in intelligent systems (Castillo and Melin 2008). To handle these uncertainties, Lotfi Zadeh (1975) proposed a more sophisticated kind of fuzzy sets theory that is called type-2 fuzzy sets (Mizumoto and Tanaka 1976; Mendel 2001).

2.2.2 Interval Type-2 Fuzzy Sets

The concept of a type-2 fuzzy set was introduced first by Zadeh (1975) as an extension of the type-1 fuzzy set. In particular, the membership function of a general type-2 fuzzy set is three-dimensional (Fig. 2.4):

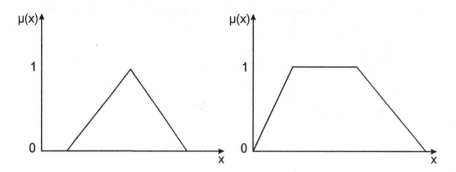

Fig. 2.3 Examples of type-1 fuzzy sets

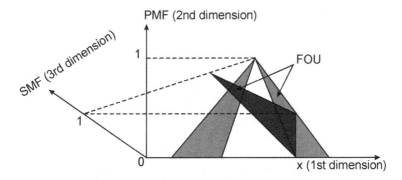

Fig. 2.4 The membership function of a general type-2 fuzzy set

- **1st dimension**: the primary variable x (e.g. age, height, grade, temperature)
- **2nd dimension**: the primary membership function (PMF), which is a function and not just a value between 0 and 1.
- **3rd dimension**: the secondary membership function (SMF), which is the value of the membership function at each point on its two-dimensional domain that is called its footprint of uncertainty (FOU). The value of SMF is, also, range between 0 and 1.

Using type-2 fuzzy logic can reduce the amount of uncertainty in a system. This is happened due to the fact that type-2 fuzzy logic offers better capabilities to handle linguistic uncertainties by modeling vagueness and unreliability of information (Liang and Mendel 2000). Such sets are useful in circumstances where it is difficult to determine the exact membership function for a fuzzy set, as in modeling a word by a fuzzy set.

When the value of the third dimension is the same (e.g. 1) everywhere, then the type-2 fuzzy set is called interval type-2 fuzzy set. For an interval type-2 set the SMF is ignored and only the FOU is used to describe it. The more (less) area in the FOU the more (less) is the uncertainty (Mendel 2001). The FOU represents the blurring of a type-1 membership function. It is completely described by its two bounding functions (Fig. 2.5): (i) a lower membership function (LMF) and (ii) an upper membership function (UMF).

2.2.3 Rule-Based Fuzzy Logic System

Type-2 fuzzy sets are finding very wide applicability in rule-based fuzzy logic systems (FLSs). The operation of FLSs is based on rules. The rules are expressed as a collection of IF-THEN statements (e.g. If George's grade at mathematics is 65/100, then he is classified to moderate students). Fuzzy sets are associated with the terms that appear in the antecedents (IF-part) or consequents (THEN-part) of rules. For example in the example "if George's grade at mathematics is 65/100, then he is classified to moderate students", the fuzzy set 'moderate' appears in

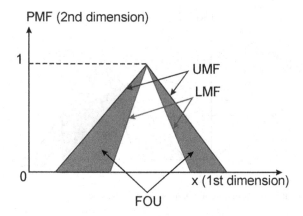

Fig. 2.5 The membership function of an interval type-2 fuzzy set

the consequents, while in the example "if the temperature indicates cold, then the heater must be switched on", the fuzzy set 'cold' appears in the antecedents. Membership functions are used to describe these fuzzy sets.

Experts construct the rules of a FLS considering their experience or data that have been extracted from experiments or surveys. Therefore, the knowledge and data that are used to construct the rules of a FLS are uncertain. This uncertainty leads to rules that have uncertain antecedents and/or consequents, which in turn translates into uncertain corresponding membership functions (Karnik et al. 1999). This uncertainty can be handled using type-2 fuzzy sets.

A type-2 FLS is depicted in Fig. 2.6. Two steps are required to go from an interval type-2 fuzzy set to a number:

- **Type-reduction**: in this step an interval type-2 fuzzy set is reduced to an interval-valued type-1 fuzzy set. This is achieved using particular algorithms. There are a comparable number of type-reduction methods (Mendel 2001).
- **Defuzzification**: In this step the centroid of the type-reduced set is computed. In particular, the average of the two end-points of the finite interval of numbers, which has been come off the process of type-reduction, is calculated. In other words, defuzzification maps the type-1 FS that came of the type-reduction step.

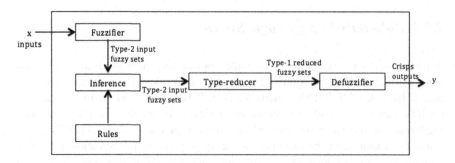

Fig. 2.6 A type-2 rule-based fuzzy logic system

2.2.4 Applications of Fuzzy Logic

The ability of fuzzy logic to handle the uncertainty, imprecise and incomplete data, and information that is characterized by human subjectivity makes it useful in many human-centric fields. Mendel (2007) has categorizes the applications of fuzzy logic in: approximation; clustering; control; databases; decision making; embedded agents; health care; hidden Markov models; neural networks; noise cancellation; pattern classification; quality control; spatial query; wireless communications. In addition, fuzzy set theory has been applied in education and educational systems. The applications of fuzzy logic in the educational field can be categorized into:

- **Grading systems**: Fuzzy logic is used to define the grade (as a letter, as a number, or as a percentage) that characterized the student's level of achievement. Examples of fuzzy applications in grading systems are the researches of (Bai and Chen 2006a, 2006b, 2008; Biswas 1995; Cheng and Yang 1998; Echauz and Vachtsevanos 1995; Law 1996; Wang and Chen 2006; Wilson et al. 1998).
- **Student's evaluation**: It includes an overall assessment of the student's learning. In particular, it is a complex process that includes student's performance, abilities, skills and learning characteristics. Some of the fuzzy logic applications in the process of the student's evaluation, which appear in the literature, are the following: (Chang and Sun 1993; Chen and Lee 1999; Ma and Zhou 2000; Nykänen 2006; Weon and Kim 2001).
- **Learning adaptation**: Learning and teaching are complex processes that have to consider each individual student's characteristics and abilities in order to be effective. The educational systems have to adapt dynamically to each individual learner's needs and abilities. Many researchers (Alves et al. 2008; Jili et al. 2009; Jurado et al. 2008; Kosba et al. 2003; Suarez-Cansino and Hernandez-Gomez 2008) have used fuzzy logic for providing learning and teaching adaptation.

2.2.4.1 Applications of Fuzzy Logic in Student Modeling

The aim of the adaptive and/or personalized tutoring systems is to readjust each time the instructional process and the teaching strategy considering the student's needs and abilities. This operation is based on human subjectivity and conceptualizations. That is the reason for the need of fuzzy logic. Therefore, there are many researchers that have used fuzzy logic techniques in student modeling to deal with uncertainty in the student's diagnose. For example, Xu et al. (2002) have used fuzzy models to represent a student profile in order to provide personalized learning materials, quiz and advices to each student. Furthermore, Kavčič (2004a) have succeeded to provide personalization of navigation in the educational content of InterMediActor system through the construction of a navigation graph and the adoption of fuzzy logic into student reasoning. A fuzzy-based student model has been applied, also, by Stathacopoulou et al. (2005) to a discovery-learning environment that aimed to help students to construct the concepts of vectors in physics and mathematics. The particular fuzzy-based student model allows the diagnostic model to some extent imitate teachers in diagnostic students' characteristics, and equips the intelligent learning environment with reasoning capabilities that can be further used to drive pedagogical decisions depending on the student learning style. Moreover, Jia et al. (2010) have applied fuzzy set theory to the design of an adaptive learning system in order to help learners to memory the content and improve their comprehension. Also, Goel et al. (2012) have used a fuzzy student model for facilitating the student reasoning process, which is based on imprecise information coming from the student-computer interaction, and predicting the degree of error that a student is possible to make in the next attempt to a problem. In addition, Salim and Haron (2006) have provided a personalized learning environment that exploit pedagogical model and fuzzy logic techniques. Other educational systems that have incorporated fuzzy logic techniques into the student model are: F-CBR-DHTS (Tsaganoua et al. 2003); TADV (Kosba et al. 2003, 2005) and DEPTHS (Jeremić et al. 2012).

2.3 Fuzzy Logic for Knowledge Representation

The knowledge domain module is one of the most major modules of an Intelligent Tutoring System (ITS). The knowledge domain representation is the base for the representation of the learner's knowledge, which is usually performed as a subset of the knowledge domain. It contains a description of the knowledge or behaviors that represent expertise in the subject-matter domain the ITS is teaching. In other words, the knowledge domain module is responsible for the representation of the subject matter taking into account the course modules, which involve domain concepts. The particular module has been introduced in ITS but its use has been extended to most current educational software applications that aim to be adaptive and/or personalized.

To enable communication between system and learner at content level, the domain model of the system has to be adequate with respect to inferences and relations of domain entities with the mental domain of a human expert (Peylo et al. 2000). Therefore, the knowledge domain representation in an adaptive and/or personalized tutoring system is an important factor for providing adaptivity. The appropriate approach for knowledge representation makes easier the selection of the appropriate educational material satisfying the student's learning needs. The most common used techniques of knowledge domain representation in adaptive tutoring systems are hierarchies and networks of concepts.

A hierarchical knowledge representation is usually used in order to specify the order in which the domain concepts of the learning material have to be taught (Chen and Shen 2011; Siddara and Manjunath 2007; Vasandani and Govindury 1995), and can be implemented through trees (Kumar 2005; Geng et al. 2011). For example, in INMA, which is a knowledge-based authoring tool for music education, the knowledge domain is described in terms of hierarchies (Virvou et al. 2006). Also, Siddappa et al. (2009) have developed a multilevel hierarchical model for the representation of knowledge domain of an intelligent tutoring system for numerical method (ITNM). This multilevel hierarchical model was based on various aptitude levels of students. An example of hierarchical representation is depicted in Fig. 2.7.

Hierarchies give information about the order in which the learning material should be taught, but they do not clearly depict the relations among the domain concepts. The network of concepts gives this kind of information. In a network of concepts, nodes represent concepts and arcs represent relations between concepts (Fig. 2.8). Many adaptive tutoring systems, such as Web-PTV (Tsiriga and Virvou 2003a, 2003b), DEPTHS (Jeremić et al. 2009) and IDEAL (Khamis 2011) use a network of concepts for representing the knowledge domain. However, in a network of concepts the relations between concepts are restricted to "part-of", "is-a" and prerequisite relations. They do not depict how the knowledge of a domain concept may be affected by the knowledge of another concept. They do not give answers to the questions: "If a student learns the concept C_i, which will be her/his knowledge level of the depended domain concept C_j?"; "If the student's knowledge of concepts C_i improves, how will be affected her/his knowledge of the

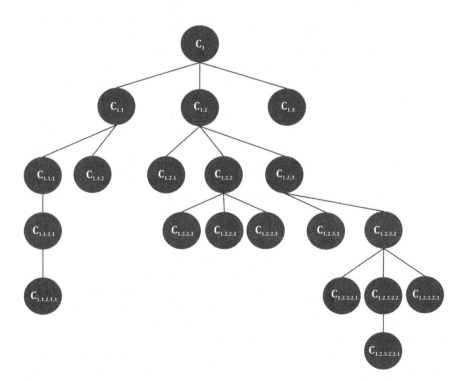

Fig. 2.7   A hierarhical tree

Fig. 2.8 A network of
concepts

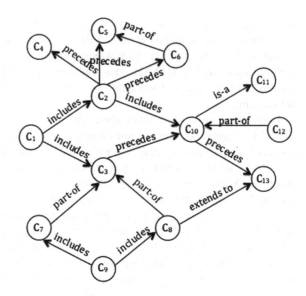

depended concept C_j?"; "If the student has misconceptions on the domain concept
C_i, how will be affected her/his knowledge level of the depended concept C_i?".

The domain concepts that constitute the learning material are not independent
from each other. The student's knowledge level of a domain concept usually is
affected by her/his knowledge level of other related domain concepts. For example,
a new domain concept may be completely unknown to the learner but in other
circumstances it may be partly known due to previous related knowledge of the
learner. On the other hand, domain concepts, which were previously known by
the learner, may be completely or partly forgotten. Hence, currently they may be
partly known or completely unknown. Therefore, the knowledge representation
approach has to allow the system to recognize either the domain concepts that
are already partly or completely known for a learner, or the domain concepts that
s/he has forgot, taking into account the learner's knowledge level of the related
concepts. Therefore, the representation of dependencies between the domain concepts of the learning material includes imprecise and uncertain information. As a
result an effective solution for handling this uncertainty is to use fuzzy logic techniques in the representation of the knowledge domain.

A fuzzy logic application, which is used to model the behavior of complex systems (Leon et al. 2011) and emphasizes the connections and dependencies between
the system's elements, is the Fuzzy Cognitive Map (FCM). Fuzzy Cognitive
Maps (FCMs) constitute a way to represent real-world dynamic systems; in a
form that corresponds closely to the way humans perceive it (Papageorgiou 2011;
Papageorgiou and Iakovidis 2013). They are able to incorporate experts' knowledge (Papageorgiou and Salmeron 2012; Salmeron 2009; Salmeron et al. 2012)
and approach representation of knowledge by emphasizing the connections and
the structure (Lin 2007). A FCM illustrates the whole system as a combination of

concepts and the various relations that exist between its concepts (Azadeh et al. 2012; Song et al. 2011; Stula et al. 2010). They are inference networks, using cyclic directed graphs, for knowledge representation and reasoning (Fig. 2.9). In particular, A FCM consists of nodes (N_1, N_2, ... N_n), which represent the important elements of the mapped system, and directed arcs, which represent the causal relationships between two nodes (N_i, N_j). The directed arcs are labeled with fuzzy values (f_{ij}) in the interval $[-1, 1]$ that show the "strength of impact" of node N_i on node N_j. If f_{ij} has a positive value, then it indicates that node N_i affects positively node N_j. In other words, the positive value on the directed arc that connects N_i with N_j, means that the increase of the value of N_i leads to the increase of the value of N_j, or the decrease of the value of N_i leads to the decrease of the value of N_j. Otherwise, If f_{ij} has a negative value, then it indicates that node N_i affects negatively node N_j. In other words, the negative value on the directed arc that connects N_i with N_j, means that the increase of the value of N_i leads to the decrease of the value of N_j, or the decrease of the value of N_i leads to the increase of the value of N_j. Therefore, a FCM is a cognitive map whose relations between the nodes can be used to compute the "strength of impact" of these elements. This property of FCM makes it able to predict, to make decisions, to generate a more accurate description of a difficult situation and to explain behaviors, actions and situations (Codara 1998). That is the reason of their extensive use in a wide range of applications (Craiger et al. 1996; Kosko 1999; Miao and Liu 2000; Rodriguez-Repiso et al. 2007; Stylios and Groumpos 2004). Furthermore, according to Papageorgiou (2011), in the past decade, FCMs have gained considerable research interest and are widely used to analyze causal systems such as system control, decision-making, management, risk analysis, text categorization, prediction etc. However, the contribution of FCMs to the knowledge representation of an adaptive tutoring system has not been discussed before.

Taking into account the above, there is the need to represent the knowledge dependency relations between the individual domain concepts of the domain knowledge. In particular, the knowledge dependencies that exist between the domain concepts of the learning material, as well as their "strength of impact" on each other have to be represented. A solution to this is to use a combination of a network of concepts with Fuzzy Cognitive Maps. In this way, a new approach of domain knowledge representation derives. That new approach is called Fuzzy Related-Concept Network (FR-CN).

Fig. 2.9 A fuzzy cognitive map

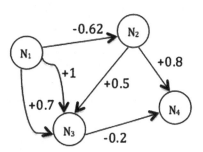

2.3.1 Knowledge Domain Representation Using a Fuzzy Related-Concept Network

A Fuzzy Related-Concepts Network is a network of concepts, which depicts, also, the knowledge dependencies that exist between the domain concepts of the learning material. Therefore, it illustrates so the structure of the learning material, as the concepts' knowledge dependencies. Particularly, it represents the fact that the knowledge level of a domain concept is increased when the knowledge level of a related topic improves, as well as the fact that the knowledge level of a domain concept is decreased when the knowledge level of a depended topic is not satisfactory. The Fuzzy Related-Concepts Network (Fig. 2.10) consists of: nodes, which depict the domain concepts of the learning material, and directed arcs, which represent relations between the concepts of the learning material.

The relations that exist between the concepts of the learning material depict so the order in which the domain concepts have to be delivered and the structure of the learning material, as the knowledge dependencies. In particular, there are three type of relations between the concepts: "precedes" that declares the order in which each domain concept of the learning material has to be taught

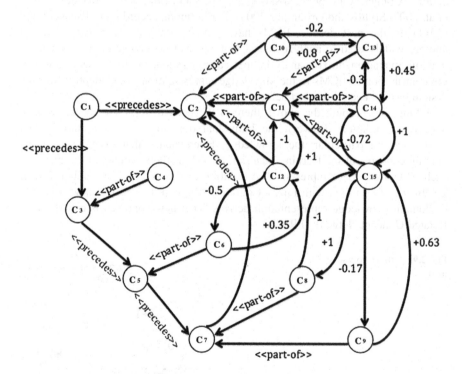

Fig. 2.10 A fuzzy related-concepts network

(for example, in Fig. 2.10 the domain concept C_3 is delivered to the learner before the domain concept C_5); "part-of" that declares that a concept belongs to another concept (for example, in Fig. 2.10 the domain concept C_2 in includes the domain concepts C_{10}, C_{11} and C_{12}); the dependence relation that declares that the knowledge level of a domain concept is affected by the learner's knowledge level on another related concept (For example, in Fig. 2.10 the knowledge level of the domain concepts C_{14}, C_8 and C_9 is affected by the learner's knowledge level on the concept C_{15}).

The dependence relations allow the tutoring system to identify how the knowledge level of a concept is affected by the learner's knowledge level on other related concepts. A dependence relation is characterized by the symbol '+' or the symbol '−' and a number (strength of impact). The symbol depicts the order in which the two related concepts are delivered to the learner. If the symbol '+' is labeled on the arc that connects C_i with C_j with direction from C_i to C_j ($C_i \rightarrow C_j$), then it denotes that C_i is taught before C_j. Otherwise, if the symbol that is labeled on the particular directed arc is the symbol '−', then it denoted that that C_j is taught before C_i. The numbers that are labeled on the directed arcs depict the degree at which the knowledge level of a domain concept is affected regarding the knowledge level of its related domain concepts. In other words, they depict the "strength of impact" of a domain concept on a related concept. The particular numbers are only positive. This is happened due to the fact that the increase of the knowledge level of a domain concept leads to the increase of the knowledge level of a depended domain concept, and the decrease of the knowledge level of a domain concept leads to the decrease of the knowledge level of a depended domain concept. Therefore, the numbers of the directed arcs that depict the knowledge dependencies belong to the interval (0, 1]. For example, in Fig. 2.10, the value '+0.8' that is labeled on the directed arc, which connects C_{10} with C_{13} ($C_{10} \rightarrow C_{13}$), denotes that the concept C_{10} is delivered to the learner before the concept C_{13} and the "strength of impact" of C_{10} on C_{13} is 0.8. Similarly, the value '−0.72' that is labeled on the directed arc, which connects C_{15} with C_{14} ($C_{15} \rightarrow C_{14}$), denotes that the concept C_{15} is delivered to the learner after the concept C_{14} and the "strength of impact" of C_{15} to C_{14} is 0.72.

The arcs in the FR-CN, which represent the domain concepts' dependencies of the knowledge domain, are bidirectional. Furthermore, the value of the arc $C_i \rightarrow C_j$ is not essentially equal to the value of the arc $C_j \rightarrow C_i$. This is happened due to the fact that changes on the knowledge level of C_i may affect the knowledge level of C_j in a different degree than changes on the knowledge level of C_j affect the knowledge level of C_i. It has to be clear that the value 1 on the directed arc that connects two dependent domain concepts does not mean that the two dependent concepts are the same. It implies that if a learner knows a domain concept of a section, s/he may know a related concept of another section at the same degree. The percentage of increase or decrease of the knowledge level of a domain concept that occurs due to changes on the knowledge level of another concept related with this domain concept is defined by experts of the knowledge domain.

Therefore, a FR-CN that is used to represent the knowledge domain of the learning material is a 6-tuple (C, ORD, PART, IMPACT, KL, f), where:

- C = {C_1, C_2, ... C_n} is the set of concepts of the knowledge domain.
- ORD: (C_i, C_j) → {0, 1} is a matrix, which denotes that the concept C_i is delivered to the learner before the concept C_j (the value of the corresponding matrix's cell—line i, column j—is 1). If the value of the corresponding matrix's cell is 0, then it denotes that there is no "precedes" relation between the two domain concepts.
- PART: (C_i, C_j) → {0, 1} is a matrix, which denotes that the concept C_i is part-of the concept C_j (the value of the corresponding matrix's cell—line i, column j—is 1). If the value of the corresponding matrix's cell is 0, then it denotes that there is no "part-of" relation between the two domain concepts.
- IMPACT: (C_i, C_j) → w_{ij} is a matrix, where w_{ij} is a weight of the directed arc from C_i to C_j, which denotes the "strength of impact" of the concept C_i on the concept C_j (the value w_{ij} is inserted in the cell that corresponds to line i and column j). If $w_{ij} = 0$, then it denotes that C_i and C_j are not knowledge related concepts.
- KL is a function that at each concept C_i associates the sequence of its activation degree. In other worlds, $KL_i(t)$ indicates the value of a concept's knowledge level at the moment t.
- f is a transformation function. For the definition of the transformation function the following limitation has to be taken into account. Only the knowledge level of the most recently read concept affects the knowledge level of a domain concept, each time. The reason for this is the fact that the learner's knowledge level is affected either by the new knowledge that s/he has obtained, or by the knowledge that s/he has forgot, each time. Consequently, the KL value of a concept is affected only by the KL value of the most recently read concept, regarding the weight of the directed arc that connects them. Therefore, the transformation function for a FR-CN, which is used to represent the knowledge domain of the learning material, is defined as: $KL_i(t + 1) = f(KL_i(t) \pm w_{ji}*p_j*KL_i(t)/100)$, where p_j is the percentage of the difference on the value of the knowledge level of the most recently read concept C_j, with $p_j = (KL_j(t + 1) - KL_j(t))*100/KL_j(t)$. Also, the + is used in case of increase and the − is used in case of decrease.

For example, the matrixes ORD (Table 2.1), PART (Table 2.2) and IMPACT (Table 2.3) for the FR-CN that depicts in Fig. 2.10 are the following:

At the ORD matrix the value of the cell ORD [i, j], which corresponds to the line i and column j, can be 1, although there is no a direct arc in the corresponding FR-CN that connects the node-concept C_i with the node-concept C_j and declares "precedes" relation between the particular concepts. The reason for that is the fact that an indirect relation of type "precedes" can be exist between the particular concepts. For example, in the FR-CN of Fig. 2.10, the concept C_3 precedes the concept C_2 due to the fact that the concept C_7 precedes the concept C_2 and the

Table 2.1   ORD

	C1	C2	C3	C4	C5	C6	C7	C8	C9	C10	C11	C12	C13	C14	C15
C1	0	1	1	1	1	1	1	1	1	1	1	1	1	1	1
C2	0	0	0	0	0	0	0	0	0	0	0	0	0	0	0
C3	0	1	0	0	1	1	1	1	1	1	1	1	1	1	1
C4	0	1	0	0	1	1	1	1	1	1	1	1	1	1	1
C5	0	1	0	0	0	0	1	1	1	1	1	1	1	1	1
C6	0	1	0	0	0	0	1	1	1	1	1	1	1	1	1
C7	0	1	0	0	0	0	0	0	0	1	1	1	1	1	1
C8	0	1	0	0	0	0	0	0	0	1	1	1	1	1	1
C9	0	1	0	0	0	0	0	0	0	1	1	1	1	1	1
C10	0	0	0	0	0	0	0	0	0	0	0	0	0	0	0
C11	0	0	0	0	0	0	0	0	0	0	0	0	0	0	0
C12	0	0	0	0	0	0	0	0	0	0	0	0	0	0	0
C13	0	0	0	0	0	0	0	0	0	0	0	0	0	0	0
C14	0	0	0	0	0	0	0	0	0	0	0	0	0	0	0
C15	0	0	0	0	0	0	0	0	0	0	0	0	0	0	0

Table 2.2   PART

	C1	C2	C3	C4	C5	C6	C7	C8	C9	C10	C11	C12	C13	C14	C15
C1	0	0	0	0	0	0	0	0	0	0	0	0	0	0	0
C2	0	0	0	0	0	0	0	0	0	0	0	0	0	0	0
C3	0	0	0	0	0	0	0	0	0	0	0	0	0	0	0
C4	0	0	1	0	0	0	0	0	0	0	0	0	0	0	0
C5	0	0	0	0	0	0	0	0	0	0	0	0	0	0	0
C6	0	0	0	0	1	0	0	0	0	0	0	0	0	0	0
C7	0	0	0	0	0	0	0	0	0	0	0	0	0	0	0
C8	0	0	0	0	0	0	1	0	0	0	0	0	0	0	0
C9	0	0	0	0	0	0	1	0	0	0	0	0	0	0	0
C10	0	1	0	0	0	0	0	0	0	0	0	0	0	0	0
C11	0	1	0	0	0	0	0	0	0	0	0	0	0	0	0
C12	0	1	0	0	0	0	0	0	0	0	0	0	0	0	0
C13	0	1	0	0	0	0	0	0	0	0	1	0	0	0	0
C14	0	1	0	0	0	0	0	0	0	0	1	0	0	0	0
C15	0	1	0	0	0	0	0	0	0	0	1	0	0	0	0

concept C_3 precedes the concept C_7. Therefore, ORD $[3, 2] = 1$. Similarly, C_4 precedes C_8 because C_4 is part-of the concept C_3, which precedes the concept C_7 whose part is the concept C_8. C_3 precedes C_7 due to the fact that C_3 precedes C_5, which precedes C_7. As a result, ORD $[4, 8] = 1$.

Table 2.3 IMPACT

	C1	C2	C3	C4	C5	C6	C7	C8	C9	C10	C11	C12	C13	C14	C15
C1	0	0	0	0	0	0	0	0	0	0	0	0	0	0	0
C2	0	0	0	0	0	0	0	0	0	0	0	0	0	0	0
C3	0	0	0	0	0	0	0	0	0	0	0	0	0	0	0
C4	0	0	0	0	0	0	0	0	0	0	0	0	0	0	0
C5	0	0	0	0	0	0	0	0	0	0	0	0	0	0	0
C6	0	0	0	0	0	0	0	0	0	0	0	+0.35	0	0	0
C7	0	0	0	0	0	0	0	0	0	0	0	0	0	0	0
C8	0	0	0	0	0	0	0	0	0	0	0	0	0	0	+1
C9	0	0	0	0	0	0	0	0	0	0	0	0	0	0	+0.63
C10	0	0	0	0	0	0	0	0	0	0	0	0	+0.8	0	0
C11	0	0	0	0	0	0	0	0	0	0	0	+1	0	0	0
C12	0	0	0	0	0	-0.5	0	0	0	0	-1	0	0	0	0
C13	0	0	0	0	0	0	0	0	0	-0.2	0	0	0	+0.45	0
C14	0	0	0	0	0	0	0	0	0	0	0	0	-0.3	0	+1
C15	0	0	0	0	0	0	0	-1	-0.17	0	0	0	0	-0.72	0

2.3.1.1 Application of FR-CN for the Representation of the Knowledge Domain of the Programming Language 'C'

An application of Fuzzy Related-concepts Networks to a real situation is needed to understand the above, described approach for knowledge domain representation. That is the aim of the particular section, in which the description of the knowledge domain of a programming tutoring system is presented. In particular, the knowledge domain of the programming tutoring system is the programming language 'C'. The aim of the particular tutoring system is to teach learners so the principles and structures of the programming language 'C', as the logic of programming. So, the learning material includes not only expressions, operations and statements of the programming language 'C', but also it includes algorithms, like calculating sums, averages and maximums or minimums. Thereby, the learning material is decomposed in domain concepts which concern declarations of variables and constants, expressions and operators, input and output expressions, the sequential execution of a program, the if, if-else and if-else if statements, the iteration statements (for loop, while loop, do...while loop), sorting and searching algorithms, arrays, functions (Table 2.4).

Learners of programming languages have different backgrounds and their knowledge of a concept of the programming language, which they are taught, is subject to change. A new concept may be completely unknown to the learner but in other circumstances it may be partly or completely known due to previous related knowledge of the learner. For example, if a learner already knows an algorithm (e.g., calculating the sum of integers in a 'for' loop), there is no need to learn another similar algorithm (e.g., counting in a 'for' loop). Similarly, if a learner knows a programming structure (e.g., one-dimensional arrays), it is easier to understand another

Table 2.4 Learning material of the programming language 'C'

C_1. Basics	$C_{1.1}$. Constants and variables	C_5. Iteration structure Unknown no of loops	$C_{5.1}$. While statement
	$C_{1.2}$. Assignment statement		$C_{5.2}$. Calculating sum in a while loop
	$C_{1.3}$. Arithmetical operators		$C_{5.3}$. Counting in a while loop
	$C_{1.4}$. Comparative operators		$C_{5.4}$. Calculating avgr in a while loop
	$C_{1.5}$. Logical operators		$C_{5.5}$. Calculating max/min in a while loop
	$C_{1.6}$. Mathematical functions		$C_{5.6}$. Do…while statement
	$C_{1.7}$. Input-output statements		
C_2. Sequence structure	$C_{2.1}$. A simple program structure	C_6. Arrays	$C_{6.1}$ One-dimensional arrays
			$C_{6.2}$. Searching
C_3. Conditional structures	$C_{3.1}$. If statement		$C_{6.3}$. Sorting
	$C_{3.2}$. If…else if		$C_{6.4}$. Two-dimensional arrays
	$C_{3.2.1}$ Methodology of finding max/min		
	$C_{3.3}$. Nested if		$C_{6.5}$. Processing per row
			$C_{6.6}$. Processing per column
C_4. Iteration structure Concrete no of loops	$C_{4.1}$. For statement		$C_{6.7}$. Processing of diagonals
	$C_{4.2}$. Calculating sum in a for loop	C_7. Sub-programming	$C_{7.1}$. Functions
	$C_{4.3}$. Counting in a for loop		
	$C_{4.4}$. Calculating avgr in a for loop		
	$C_{4.5}$. Calculating max/min in a for loop		

programming structure (e.g., multidimensional arrays), so this new structure should not be considered as being completely unknown to the learner. On the other hand, domain concepts, which were previously known by the learner, may be completely or partly forgotten. For example, if a learner has difficulties in calculating a sum in a 'while' loop, her/his knowledge of the previous domain concept of "calculating a sum in a 'for' loop" has eroded. Therefore, there is the need to represent the knowledge dependencies that exist between the domain concepts of the learning material of the programming language. This is achieved using Fuzzy Related-Concepts Network. The FR-CN for the knowledge domain of the programming language 'C' that is described in Table 1.7 is depicted in Fig. 2.11. Tables 2.5, 2.6 and 2.7 are a part of the matrixes ORD, PART, IMPACT of the FR-CN of Fig. 2.11 correspondingly. The whole matrixes are presented in the Appendix A.

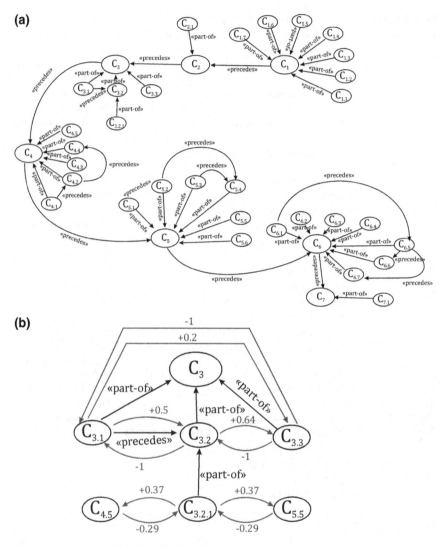

Fig. 2.11 The FR-CN of the knowledge domain of the programming language 'C' (it is decomposed in four graphs). **a** The "precedence" and "part-of" relations of the FR-CN; **b** The knowledge dependence relations for the domain concepts of the section 3; **c** The knowledge dependence relations for the domain concepts of the section 6; **d** The knowledge dependence relations for the domain concepts of the sections 4 and 5

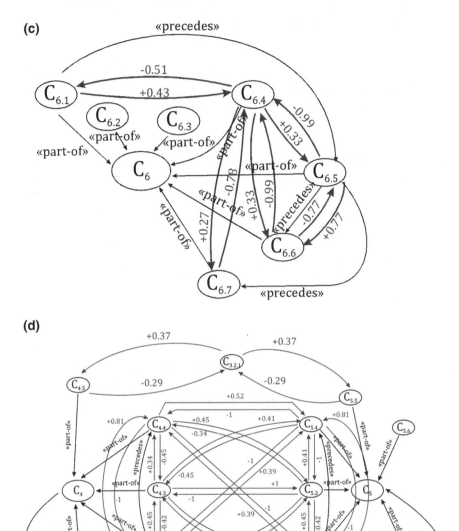

Fig. 2.11 (continued)

The value 1 on the directed arc that connects two dependent domain concepts of the FR-CN implies that if a learner knows a domain concept, then s/he may know a related domain concept at the same degree. For example, if a learner has been tested and found to have known the "for" loop and the "while" loop and this

Table 2.5 A sample of the ORDER matrix of the FR-CN of Fig. 2.11

	C_1	$C_{1.1}$	$C_{1.2}$	$C_{1.3}$	$C_{1.4}$	$C_{1.5}$	$C_{1.6}$	$C_{1.7}$	C_2	$C_{2.1}$	C_3	$C_{3.1}$	$C_{3.2}$	$C_{3.2.1}$	$C_{3.3}$
$C_{1.7}$	0	0	0	0	0	0	0	0	1	1	1	1	1	1	1
C_2	0	0	0	0	0	0	0	0	0	0	1	1	1	1	1
$C_{2.1}$	0	0	0	0	0	0	0	0	0	0	1	1	1	1	1
C_3	0	0	0	0	0	0	0	0	0	0	0	0	0	0	0
$C_{3.1}$	0	0	0	0	0	0	0	0	0	0	0	0	1	0	0

Table 2.6 A sample of the PART matrix of the FR-CN of Fig. 2.11

	C_1	$C_{1.1}$	$C_{1.2}$	$C_{1.3}$	$C_{1.4}$	$C_{1.5}$	$C_{1.6}$	$C_{1.7}$	C_2	$C_{2.1}$	C_3	$C_{3.1}$	$C_{3.2}$	$C_{3.2.1}$	$C_{3.3}$
$C_{2.1}$	0	0	0	0	0	0	0	0	1	0	0	0	0	0	0
C_3	0	0	0	0	0	0	0	0	0	0	0	0	0	0	0
$C_{3.1}$	0	0	0	0	0	0	0	0	0	0	1	0	0	0	0
$C_{3.2}$	0	0	0	0	0	0	0	0	0	0	1	0	0	0	0
$C_{3.2.1}$	0	0	0	0	0	0	0	0	0	0	1	0	1	0	0

Table 2.7 A sample of the IMPACT matrix of the FR-CN of Fig. 2.11

	C_4	$C_{4.1}$	$C_{4.2}$	$C_{4.3}$	$C_{4.4}$	$C_{4.5}$	C_5	$C_{5.1}$	$C_{5.2}$	$C_{5.3}$	$C_{5.4}$	$C_{5.5}$	$C_{5.6}$
$C_{4.1}$	0	0	0	0	0	0	0	0	0	0	0	0	0
$C_{4.2}$	0	0	0	+0.45	+0.81	0	0	0	+1	+0.45	+0.39	0	0
$C_{4.3}$	0	0	−0.42	0	+0.34	0	0	0	+0.42	+1	+0.41	0	0
$C_{4.4}$	0	0	−1	−0.45	0	0	0	0	+1	+0.45	+0.52	0	0
$C_{4.5}$	0	0	0	0	0	0	0	0	0	0	0	+1	0

learner knows how to calculate sum in a "for" loop, s/he will also know how to calculate sum in a "while" loop, since the methodology is the same.

Experts on programming have defined so the domain concepts of the learning material, as their relations ("precedence", "part-of", "knowledge dependence"). In particular, ten professors of computer programming, whose experience counts 12 years at least, are responsible for the definition and structure of the knowledge domain. They were, also, asked to determine, empirically, the knowledge dependencies that exist between the defined domain concepts of the learning material, as well as their "strength of impact" on each other. The FR-CN that is depicted in Fig. 2.11 has been mapped according to the mean of the experts' answers (due to its complexity, it has been decomposed in four graphs).

The information that is derived from the above matrixes concerns:

- The order in which the domain concepts of the leaning material have to be delivered.
- Which domain concepts belong to another general domain concept of the learning material.
- The knowledge dependencies that exist between the domain concepts of the learning material and their "strength of impact".

For example, the domain concept C_1 is delivered before concept C_2 and concept $C_{4.2}$ is delivered before the domain concept $C_{4.4}$. That is derived from the values of the cells ORDER [1, 9] (Table 1.8a) and ORDER [18, 20] (Table 1.8b), which are 1 both. On the other hand, the ORDER [18, 21] $= 0$ (Table 1.8b) denotes that the concept $C_{4.2}$ is not necessary to be taught before the concept $C_{4.5}$. Furthermore, $C_{3.2.1}$ belongs to the concepts C_3 and $C_{3.2}$ as PART [14, 11] $= 1$ and PART [14, 13] $= 1$ (Table 2.1a). In addition, the learner's knowledge level on the concept $C_{4.4}$ affects the particular learner's knowledge level on the previously delivered concepts $C_{4.2}$, $C_{4.3}$, $C_{5.2}$, $C_{5.3}$ and $C_{5.5}$. This information is derived from the matrix IMPACT. In particular, the values IMPACT [20, 18] $= -1$ and IMPACT [20, 19] $= -0.45$ (Table 2.2b) denote that the knowledge level of concept $C_{4.4}$ affects the knowledge level of $C_{4.2}$ and $C_{4.3}$, and its "strength of impact" on $C_{4.2}$ and $C_{4.3}$ are 1 and 0.45 correspondingly. Similarly, the values IMPACT [20, 24] $= +1$, IMPACT [20, 25] $= +0.45$ and IMPACT [20, 26] $= +0.52$ (Table 2.2b) denote that the knowledge level of concept $C_{4.4}$ affects the knowledge level of the following concepts $C_{5.2}$, $C_{5.3}$ and $C_{5.5}$, and its "strength of impact" on the particular concepts are 1, 0.45 and 0.52 correspondingly. However, the value IMPACT [20, 21] $= 0$ (Table 2.2b) denote that the knowledge level of concept $C_{4.4}$ does not affect the knowledge level of the concept $C_{4.5}$.

2.4 A Novel Rule-Based Fuzzy Logic System for Modeling Automatically the Learning or Forgetting Process of a Student

Learning is not a "black or white" process. The definition of the learner's knowledge level is a moving target. In other words, it is not a straightforward task to define for each learner which concepts are unknown, known or assimilated and at what degree. The particular process is confronted with uncertainty and human subjectivity. One possible approach to deal with this is fuzzy set techniques, with their ability to naturally represent human conceptualization. That is the reason for the integration of fuzzy logic techniques into the student model.

Fuzzy logic is the solution for recognizing and modeling the increase and/or decrease of the learner's knowledge level on a domain concept in relation with her/his performance on other related domain concepts of the learning material. In particular, the presented rule-based fuzzy logic module is responsible for identifying and updating the student's knowledge level of all the concepts of the knowledge domain. Its operation is based on the Fuzzy Related-Concepts Network that is used to represent the structure of the learning material and the dependencies that exist between the domain concepts. It uses fuzzy sets to represent the student's knowledge level and a mechanism of rules over the fuzzy sets, which is triggered after a change has occurred on the student's knowledge level of a domain concept. This mechanism updates the student's knowledge level of all related with this concept, concepts. With this approach the alterations on the state of student's knowledge level, such as forgetting or learning are represented.

The presented rule-based fuzzy logic module includes the following three steps:

Step 1 **Definition of the fuzzy sets:**

In the particular step, the definition of the fuzzy sets, which represent the learner's knowledge level on a domain concept (i.e. {"Unknown", "Known", "Learned"} or {"Unknown", "Insufficiently Known", "Known", "Learned", "Assimilated"}), is carried out. Fuzzy sets are used to characterize the changeable learner's knowledge level. Therefore, FS_1, FS_2, ..., FS_n are the defined fuzzy sets, for the educational adaptive system.

Step 2 **Definition of the membership functions:**

In the particular step, the membership functions of the determined fuzzy sets FS_1, FS_2, ..., FS_n is defined. The membership functions (Fig. 2.12) are defined as follows (x indicates the learner's degree of success on a particular domain concept; x_{i-1}, x_i, x_{i+1}, x_{i+2} are thresholds that indicate particular degrees of success like 0, 50, 100):

$$\mu_{FS1} = \begin{cases} 1, & x \leq x_1 \\ 1 - \frac{x-x_1}{x_2-x_1}, & x_1 < x < x_2 \\ 0, & x \geq x_2 \end{cases}$$

$$\forall i \neq 1 \, and \, i \neq n \, \mu_{FSi} = \begin{cases} \frac{x-x_{2i-3}}{x_{2i-2}-x_{2i-3}}, & x_{2i-3} < x < x_{2i-2} \\ 1, & x_{2i-2} \leq x \leq x_{2i-1} \\ 1 - \frac{x-x_{2i-1}}{x_{2i}-x_{2i-1}}, & x_{2i-1} < x < x_{2i} \\ 0, & x \leq x_{2i-3} \, or \, x \geq x_{2i} \end{cases}$$

$$\mu_{FSn} = \begin{cases} \frac{x-x_{2n-3}}{x_{2n-2}-x_{2n-3}} & x_{2n-3} < x < x_{2n-2} \\ 1 & x_{2n-2} \leq x \leq x_{2n-1} \\ 0 & x \leq x_{2n-3} \end{cases}$$

The knowledge level of a domain concept changes in a continuous way. Meaning that the knowledge level of a domain concept usually passes gradually from the

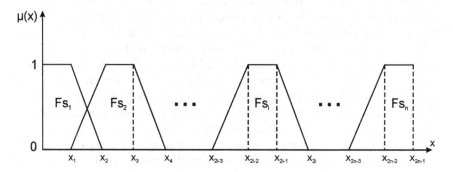

Fig. 2.12 The membership functions μ_{FSi}

unknown state to the learned and assimilated state. Membership values correspond to percentages of the offered knowledge in a way that they cover 100 % of it, at any time. This gives a more natural and understandable way of representation. For example, it would be non-intuitive to say that domain concept "A" is 0.5 (50 %) Insufficiently Known and 0.6 (60 %) Known for a student, given that 0.5 plus 0.6 gives 1.1 (110 %). So, the sum of the concept's percentage of different knowledge levels has to be 100 %, or 1 if the membership value of a concept to a knowledge level category is from 0 to 1. So, the following expression stands:

$$\mu_{FS_1} + \mu_{FS_2} + \mu_{FS_3} + \ldots + \mu_{FS_n} = 1$$

Therefore, a set (μ_{FS1}, μ_{FS2}, μ_{FS3}, … ,μ_{FSn}) is used express the student knowledge of a domain concept.

Step 3   Definition of the fuzzy rules:

When there is a dependency between two domain concepts, then the knowledge level of the one domain concept can affect the knowledge level of the other domain concept. More specifically, the following are taken into account:

- Considering the knowledge level of C_i, the knowledge level of its following domain concept C_j is increased or decreased.
- Considering the knowledge level of C_j, the knowledge level of its prerequisite domain concept C_i is increased or decreased.

Consequently, the student model expands when a change on the knowledge level of a domain concept causes increase on the knowledge level of the related concepts, or it is minimized when a change on the knowledge level of a domain concept causes decrease on the knowledge level of the related concepts with this concept.

In this document, D is defined to represent the knowledge dependency between two domain concepts. The symbolism $\mu_D(C_i, C_j)$ is used to represent the "strength of impact" of C_j on C_i and the symbolism $\mu_D(C_j, C_i)$ is used to represent the "strength of impact" of C_i on C_j. The values of $\mu_D(C_i, C_j)$ and $\mu_D(C_i, C_j)$ are the values of the arcs that depict the "knowledge dependencies" relations between the concepts of the learning material in the FR-CN of the knowledge domain (Sect. 3.1).

Concerning two domain concepts C_i and C_j where C_i is taught before C_j, the knowledge level of the concepts can change according to the following rules. These rules depict how the changes on the knowledge level of the domain concepts of the learning material for a student occur, revealing her/his learning state. In particular, they reveal if s/he learns or not or if s/he forgets. If the knowledge level of a concept is decreased, then the system infers that the student does not learn. If the knowledge level of a previously taught concept is decreased, then the system infers that the student forgets. If the knowledge level of a concept is increased, then the system infers that the student learns, and if the knowledge level of all the related concepts is improved continuously, then the system infers that the student assimilates the learning material.

The rules are based on Kavčič's (2004b) work. That work models mainly how the student's knowledge level of the prerequisites concepts that the student had read previously, is improved when s/he performs better in following concepts. In this way Kavčič's work deals only with how learning progresses. In her work there are no rules that imply the possible decrease of knowledge via the student's forgetting of some previously learned concepts. Moreover, another important problem that is not dealt with in Kavčič's work is the fact that in static educational systems, students are often required to repeat previously known concepts thought the following chapters. However, this practice is quite generic and does not take into account individual features of a student such as how fast they learn or how well they remember previously taught concepts. As such, educational systems do not adapt their pace on individual students. In view of the above, in the presented rule-based fuzzy module, Kavčič's rules have been expanded to deal with the above problems. The rules with these novelties that lead to the dynamic personalization of teaching are presented below. In the following rules, FS_x, FS_y are fuzzy sets that represent knowledge levels with $FS_x < FS_y$, and $KL()$ denotes the "Knowledge Level of".

- **Based on updates of the $KL(C_i)$, the $KL(C_j)$ is improved according to:**

  R1: If the same fuzzy sets are active for both C_i and C_j, then $KL(C_j) = FS_x$ with

  $$\mu_{FS_y}(C_j) = \max[\mu_{FS_x}(C_i), \mu_{FS_x}(C_i) * \mu_D(C_i, C_j)$$

  where FS_x is the last active fuzzy set. Subtract the value (new $\mu_{FSx}(C_j)$—previous $\mu_{FSx}(C_j)$) from the others $\mu_{FSy}(C_j)$ ($FSy < FSx$) sequentially until $\sum \mu_{FSi} = 1$.

  R2: If $KL(C_j) = FS_x$ and $KL(C_i) = FS_y$, then $KL(C_j) = FS_y$ with

  $$\mu_{FS_y}(C_j) = \mu_{FS_y}(C_i) * \mu_D(C_i, C_j)$$

- **Based on updates of the $KL(C_i)$, the $KL(C_j)$ is deteriorated according to:**

  R3: If $KL(Cj) = FS_n$, then
  if $\mu_{FS1}(C_j) + \mu_{FS2}(C_j) + \cdots + \mu_{FSn-1}(C_j) < \mu_{FSi}(C_i) * \mu_D(C_i, C_j)$, where $i < n$, then the corresponding value is subtracted by $\mu_{FSn}(Cj)$
  else it does not change.

  R4: If $KL(C_j) = FS_y$ and $KL(C_i) = FS_x$, then $KL(C_j) = FS_x$ with

  $$\mu_{FS_x}(C_j) = \mu_{FS_x}(C_i) * \mu_D(C_i, C_j)$$

- **Based on updates of the $KL(C_j)$, the $KL(C_i)$ is improved according to:**

  R5: If the same fuzzy sets are active for both C_i and C_j, then $KL(C_j) = FS_x$ with

  $$\mu_{FS_x}(C_i) = \max[\mu_{FS_x}(C_i), \mu_{FS_x}(C_j) * \mu_D(C_i, C_j)]$$

  where FS_x is the last active fuzzy set. Subtract the value (new $\mu_{FSx}(C_i)$—previous $\mu_{FSx}(C_i)$) from the others $\mu_{FSy}(C_i)$ ($FSy < FSx$) sequentially until $\sum \mu_{FSi} = 1$

  R6: If $KL(C_i) = FS_x$ and $KL(C_j) = FS_y$, then $KL(C_i) = FS_y$ with

  $$\mu_{FS_y}(C_i) = \mu_{FS_y}(C_j) * \mu_D(C_j, C_i)$$

- **Based on updates of the KL(C_j), the KL(C_i) is deteriorated according to:**

  R7: *If KL(Ci) = FS_n with* μ_{FSn}(Ci) = 1, *then it does not change*

  R8:   The   formula $x_i = \left(1 - \mu_D\left(C_i, C_j\right)\right) * x_i + \min[\mu_D\left(C_i, C_j\right) * x_i, \mu_D\left(C_i, C_j\right) * x_j]$, where x_i and x_j are the values of the criterion, which determines the fuzzy sets that are active each time for C_i and C_j respectively, is used (for the calculation of previous x_i, the membership value of the upper active fuzzy set is used). Then, using the new x_i, the KL(C_i) is determined, calculating the membership functions.
- **Limitation:** $\sum \mu_{FSi} = 1$

2.4.1   Integration of the Fuzzy Rules

The application of the fuzzy rules of the step 3 that was described above deals with the problem of estimating wrongly the knowledge level of a domain concept. In particular, consider the fuzzy sets {"Uknown", "Known", "Well-Known", "Learned"} and the set of their membership functions (μ_{Un}, μ_K, μ_{WK}, μ_L) that represent the student's knowledge level of a domain concept. Let's the domain concept C_i to be 100 % 'Learned' and the "strength of impact" of C_i on the following concept C_j to be 0.3. The knowledge level of C_j is 100 % 'Unknown'. According to the rule R2, the knowledge level of C_j will become 30 % 'Learned'. However, that it means that the rest 70 % of the concept C_j is 'Known'? The answer is no. The rest 70 % of the C_j can be 'Unknown', 'Known', 'Well-Known' or 'Learned', or different parts of it can belong to a different fuzzy set (i.e. 10 % 'Unknown', 20 % 'Known' and 40 % 'Well-Known'). In addition, let's the set that describes the knowledge level of the domain concept C_i to be (0.8, 0.2, 0, 0) (e.g. 80 % 'Unknown' and 20 % 'Known' → KL(C_j) = 0.2 'Known') and the "strength of impact" of C_i on its following concept C_j to be 0.6. The knowledge level of C_j is 20 % 'Learned'. According to the rule R4, the knowledge level of C_j will become 60 % 'Known'. However, that it means that the rest 40 % of the concept C_j is 'Uknown'? The answer is no. It can be any of the above fuzzy sets.

A solution to this problem is to keep data for each domain concept of the learning material concerning the different part of the particular concept that can be affected be other related concepts. In such a way, the system can be informed each time about the knowledge level of each separate part of the particular domain concept and it is able to draw conclusions about the learner's knowledge level on the overall domain concept. For example, according to the Fig. 2.10 (Sect. 3.1) the domain concept C_{12} is affected by both concepts C_{11} and C_6. Initially is KL(C_6) = KL(C_{11}) = KL(C_{12}) = 100 % 'Uknown'. During the learning process, the concept C_6 is delivered to the learner firstly. The learner's knowledge level on the particular concept becomes 20 % 'Well-Known' and 80 % 'Known' (KL(C_6) = 20 % Well-Known). According to the rule R2, the learner's knowledge level on the domain concept C_{12} will become 7 % 'Well-Known' and 28 % 'Known'. The other part, however, of C_{12} is not affected by C_6. So, its knowledge level remains 'Unknown'. Therefore, C_{12} is 7 % 'Well-Known', 28 % 'Known' and 65 % 'Unknown'. As a result, the system will advise the learner to read C_{12}. Also, according to R6, the learner's knowledge level of C_{11} will become 7 % 'Well-Known' 28 % 'Known' and 65 % 'Unknown' because C_{12} affects C_{11} with "strength of impact" 1 (Fig. 2.10). Then, the concept C_{11} is delivered to the learner. The learner's knowledge level on the particular concept becomes 40 % 'Learned' and 60 % 'Well-Known' (KL(C_{11}) = 40 % Learned). According to R2 is KL(C_{12}) = 40 % Learned (40 % 'Learned' and 60 % 'Well-Known'), due to the fact that the "strength of impact" of C_{11} on C_{12} is 1. Therefore, the system will consider that the concept the learner knows C_{12}, and it will not advise her/him to read the particular concept. In addition, C_{12} affects C_6. The "strength of impact" of the particular knowledge dependency is 0.5. Therefore, according to the rule R6, the learner's knowledge level of C_6 will become 20 % 'Learned' and 30 % 'Well-known'. However, because the previous knowledge level of C_6 was 20 % 'Well-Known' and 80 % 'Known', the system will consider that the rest 50 % of C_6 remains 'Known'. Thereby, although the learner's knowledge level on C_6 has been improved, the system will advise the learner to revise the domain concept C_6.

2.4.2 Application of the Presented Rule-Based Fuzzy Logic System in a Programming Tutoring System

In this chapter an application of the presented rule-based fuzzy logic system is described. In particular, the presented rule-based fuzzy logic system is used to model the cognitive states of learners of the programming language 'C'.

Step 1   **Definition of the fuzzy sets:**

The defined fuzzy sets are the following:

- **Unknown (Un):** the degree of success in the domain concept is from 0 to 50 %.
- **Moderate Known (MKn):** the degree of success in the domain concept is from 40 to 70 %.
- **Known (Kn):** the degree of success in the domain concept is from 60 to 80 %.
- **Learned (L):** the degree of success in the domain concept is from 75 to 90 %.
- **Assimilated (A):** the degree of success in the domain concept is from 85 to 100 %.

Step 2   **Definition of the membership functions:**

The membership functions of the fuzzy sets Un, MKn, Kn, L and A are depicted in Fig. 2.13 and are the following (x indicates the learner's degree of success on a particular domain concept):

$$\mu_{Un} = \begin{cases} 1, & x \leq 40 \\ 1 - \frac{x-40}{10}, & 40 < x < 50 \\ 0, & x \geq 50 \end{cases}$$

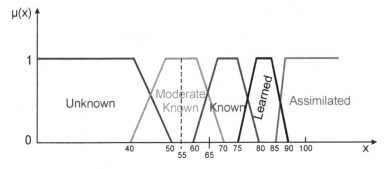

Fig. 2.13 The membership functions of the fuzzy sets of the programming tutoring system for 'C'

$$\mu_{MKn} = \begin{cases} \frac{x-40}{10}, & 40 < x < 50 \\ 1, & 50 \le x \le 60 \\ 1 - \frac{x-60}{10}, & 60 < x < 70 \\ 0, & x \le 40 \, or \, x \ge 70 \end{cases}$$

$$\mu_{Kn} = \begin{cases} \frac{x-60}{10}, & 60 < x < 70 \\ 1, & 70 \le x \le 75 \\ 1 - \frac{x-75}{5}, & 75 < x < 80 \\ 0, & x \le 60 \, or \, x \ge 80 \end{cases}$$

$$\mu_{L} = \begin{cases} \frac{x-75}{5}, & 75 < x < 80 \\ 1, & 80 \le x \le 85 \\ 1 - \frac{x-85}{5}, & 85 < x < 90 \\ 0, & x \le 75 \, or \, x \ge 90 \end{cases}$$

$$\mu_{A} = \begin{cases} \frac{x-85}{5}, & 85 < x < 90 \\ 1, & 90 \le x \le 100 \\ 0, & x \le 85 \end{cases}$$

Therefore, a set (μ_{Un}, μ_{MKn}, μ_{Kn}, μ_L, μ_A) is used to express the student knowledge of a domain concept.

Experts on programming and teachers of the programming language 'C' have defined the limits of each fuzzy set. In particular, they were asked to determine the lower and higher values of the degree of success that characterize a domain concept as 'Unknown', 'Moderate Known', 'Known', 'learned' and 'Assimilated'. The mean values of their answers consist the base for the definition of the limits of the presented fuzzy sets.

Step 3 Definition of the fuzzy rules:

Concerning two domain concepts C_i and C_j where C_i is taught before C_j, the knowledge level of the concepts can change according to the following rules ($\mu_D(C_i, C_j)$ and $\mu_D(C_j, C_i)$ indicate the "strength of impact" of C_i on C_j and of C_j on C_i correspondingly. Their values are the values of the arcs that depict the "knowledge dependencies" relations between the concepts of the learning material in the FR-CN (Sect. 3.1.1 Fig. 2.11)):

- *Based on updates of the KL(C_i), the KL(C_j) is improved according to:*

Subtract the value (new $\mu_x(C_j)$—previous $\mu_x(C_j)$) from the others $\mu_y(C_j)$ sequentially until $\mu_{Un} + \mu_{MKn} + \mu_{Kn} + \mu_L + \mu_A = 1$, where x = {MKn, Kn, L, A} and y = {Un, MKn, Kn, L} with y < x.

R1: If the same fuzzy sets are active for both C_i and C_j, then:
- If $KL_A(C_j) > 0$: $\mu_A(C_j) = \max\left[\mu_A(C_j), \mu_A(C_i) * \mu_D(C_i, C_j)\right]$
- Else If $KL_L(C_j) > 0$: $\mu_L(C_j) = \max\left[\mu_L(C_j), \mu_L(C_i) * \mu_D(C_i, C_j)\right]$
- Else If $KL_{Kn}(C_j) > 0$: $\mu_{Kn}(C_j) = \max\left[\mu_{Kn}(C_j), \mu_{Kn}(C_i) * \mu_D(C_i, C_j)\right]$
- Else If $KL_{MKn}(C_j) > 0$: $\mu_{MKn}(C_j) = \max\left[\mu_{MKn}(C_j), \mu_{MKn}(C_i) * \mu_D(C_i, C_j)\right]$

R2:

(a) If $KL(C_j) = Un$ and $KL(C_i) = MKn$, then $KL(C_j) = MKn$ with

$$\mu_{MKn}(C_j) = \mu_{MKn}(C_i) * \mu_D(C_i, C_j)$$

(b) If $KL(C_j) = Un$ and $KL(C_i) = Kn$, then $KL(C_j) = Kn$ with

$$\mu_{Kn}(C_j) = \mu_{Kn}(C_i) * \mu_D(C_i, C_j)$$

(c) If $KL(C_j) = Un$ and $KL(C_i) = L$, then $KL(C_j) = L$ with

$$\mu_L(C_j) = \mu_L(C_i) * \mu_D(C_i, C_j)$$

(d) If $KL(C_j) = Un$ and $KL(C_i) = A$, then $KL(C_j) = A$ with

$$\mu_A(C_j) = \mu_A(C_i) * \mu_D(C_i, C_j)$$

(e) If $KL(C_j) = MKn$ and $KL(C_i) = Kn$, then $KL(C_j) = Kn$ with

$$\mu_{Kn}(C_j) = \mu_{Kn}(C_i) * \mu_D(C_i, C_j)$$

(f) If $KL(C_j) = MKn$ and $KL(C_i) = L$, then $KL(C_j) = L$ with

$$\mu_L(C_j) = \mu_L(C_i) * \mu_D(C_i, C_j)$$

(g) If $KL(C_j) = MKn$ and $KL(C_i) = A$, then $KL(C_j) = A$ with

$$\mu_A(C_j) = \mu_A(C_i) * \mu_D(C_i, C_j)$$

(h) If $KL(C_j) = Kn$ and $KL(C_i) = L$, then $KL(C_j) = L$ with

$$\mu_L(C_j) = \mu_L(C_i) * \mu_D(C_i, C_j)$$

(i) If $KL(C_j) = Kn$ and $KL(C_i) = A$, then $KL(C_j) = A$ with

$$\mu_A(C_j) = \mu_A(C_i) * \mu_D(C_i, C_j)$$

(j) If $KL(C_j) = L$ and $KL(C_i) = A$, then $KL(C_j) = A$ with

$$\mu_A(C_j) = \mu_A(C_i) * \mu_D(C_i, C_j)$$

- **Based on updates of the KL(C_i), the KL(C_j) is deteriorated according to:**

R3: If $KL(Cj) = A$, then
- if $\mu_{Un}(C_j) + \mu_{MKn}(C_j) + \mu_{Kn}(C_j) + \mu_L(C_j) < \mu_x(C_i) * \mu_D(C_i, C_j)$, where $x = \{Un, MKn, Kn, L\}$, then the corresponding value is subtracted by $\mu_A(Cj)$
- else it does not change.

R4:

(a) If $KL(C_j) = L$ and $KL(C_i) = Kn$, then $KL(C_j) = Kn$ with

$$\mu_{Kn}(C_j) = \mu_{Kn}(C_i) * \mu_D(C_i, C_j)$$

(b) If $KL(C_j) = L$ and $KL(C_i) = MKn$, then $KL(C_j) = MKn$ with

$$\mu_{MKn}(C_j) = \mu_{MKn}(C_i) * \mu_D(C_i, C_j)$$

(c) If $KL(C_j) = L$ and $KL(C_i) = Un$, then $KL(C_j) = Un$ with

$$\mu_{Un}(C_j) = \mu_{Un}(C_i) * \mu_D(C_i, C_j)$$

(d) If $KL(C_j) = Kn$ and $KL(C_i) = MKn$, then $KL(C_j) = MKn$ with

$$\mu_{MKn}(C_j) = \mu_{MKn}(C_i) * \mu_D(C_i, C_j)$$

(e) If $KL(C_j) = Kn$ and $KL(C_i) = Un$, then $KL(C_j) = Un$ with

$$\mu_{Un}(C_j) = \mu_{Un}(C_i) * \mu_D(C_i, C_j)$$

(f) If $KL(C_j) = MKn$ and $KL(C_i) = Un$, then $KL(C_j) = Un$ with

$$\mu_{Un}(C_j) = \mu_{Un}(C_i) * \mu_D(C_i, C_j)$$

- **Based on updates of the KL(C_j), the KL(C_i) is improved according to:**

R5: If the same fuzzy sets are active for both Ci and Cj, then:
- If $KL_A(C_i) > 0$: $\mu_A(C_i) = \max[\mu_A(C_i), \mu_A(C_j) * \mu_D(C_j, C_i)]$
- Else If $KL_L(C_i) > 0$: $\mu_L(C_i) = \max[\mu_L(C_i), \mu_L(C_j) * \mu_D(C_j, C_i)]$
- Else If $KL_{Kn}(C_i) > 0$: $\mu_{Kn}(C_i) = \max[\mu_{Kn}(C_i), \mu_{Kn}(C_j) * \mu_D(C_j, C_i)]$
- Else If $KL_{MKn}(C_i) > 0$: $\mu_{MKn}(C_i) = \max[\mu_{MKn}(C_i), \mu_{MKn}(C_j) * \mu_D(C_j, C_i)]$

Subtract the value (new $\mu_x(C_i)$—previous $\mu_x(C_i)$) from the others $\mu_y(C_i)$ sequentially until $\mu_{Un} + \mu_{MKn} + \mu_{Kn} + \mu_L + \mu_A = 1$, where $x = \{MKn, Kn, L, A\}$ and $y = \{Un, MKn, Kn, L\}$ with $y < x$.

R6:

(a) If $KL(C_i) = Un$ and $KL(C_j) = MKn$, then $KL(C_i) = MKn$ with

$$\mu_{MKn}(C_i) = \mu_{MKn}(C_j) * \mu_D(C_j, C_i)$$

(b) If $KL(C_i) = Un$ and $KL(C_j) = Kn$, then $KL(C_i) = Kn$ with

$$\mu_{Kn}(C_i) = \mu_{Kn}(C_j) * \mu_D(C_j, C_i)$$

(c) If $KL(C_i) = Un$ and $KL(C_j) = L$, then $KL(C_i) = L$ with

$$\mu_L(C_i) = \mu_L(C_j) * \mu_D(C_j, C_i)$$

(d) If $KL(C_i) = Un$ and $KL(C_j) = A$, then $KL(C_i) = A$ with

$$\mu_A(C_i) = \mu_A(C_j) * \mu_D(C_j, C_i)$$

(e) If $KL(C_i) = MKn$ and $KL(C_j) = Kn$, then $KL(C_i) = Kn$ with

$$\mu_{Kn}(C_i) = \mu_{Kn}(C_j) * \mu_D(C_j, C_i)$$

(f) If $KL(C_i) = MKn$ and $KL(C_j) = L$, then $KL(C_i) = L$ with

$$\mu_L(C_i) = \mu_L(C_j) * \mu_D(C_j, C_i)$$

(g) If $KL(C_i) = MKn$ and $KL(C_j) = A$, then $KL(C_i) = A$ with

$$\mu_A(C_i) = \mu_A(C_j) * \mu_D(C_j, C_i)$$

(h) If $KL(C_i) = Kn$ and $KL(C_j) = L$, then $KL(C_i) = L$ with

$$\mu_L(C_i) = \mu_L(C_j) * \mu_D(C_j, C_i)$$

(i) If $KL(C_i) = Kn$ and $KL(C_j) = A$, then $KL(C_i) = A$ with

$$\mu_A(C_i) = \mu_A(C_j) * \mu_D(C_j, C_i)$$

(j) If $KL(C_i) = L$ and $KL(C_j) = A$, then $KL(C_i) = A$ with

$$\mu_A(C_i) = \mu_A(C_j) * \mu_D(C_j, C_i)$$

- *Based on updates of the KL(C_j), the KL(C_i) is deteriorated according to:*

R7: *If $KL(C_i) = A$ with* $\mu_A(Ci) = 1$, *then it does not change.*

R8: The formula $x_i = (1 - \mu_D(C_i, C_j)) * x_i + \min[\mu_D(C_i, C_j) * x_i, \mu_D(C_i, C_j) * x_j]$, where x_i and x_j are the degree of success, which determine the fuzzy sets that are active each time for C_i and C_j respectively, is used (for the calculation of previous x_i, the membership value of the upper active fuzzy set is used). Then, using the new x_i, the $KL(C_i)$ is determined, calculating the membership functions.

- *Limitation* **L1:** $\mu_{Un} + \mu_{MKn} + \mu_{Kn} + \mu_L + \mu_A = 1$.

2.4.2.1 Examples of Operation

The above described rule-based fuzzy logic system was used in a postgraduate program in the field of informatics at the University of Piraeus in Greece. It was used in order to offer dynamically personalized e-training in computer programming and the language C. At the beginning, all the domain concepts of the learning material were considered to be 'Unknown' for the learners. At the next interactions, the system delivered to them the appropriate learning material for each individual student's needs by adapting instantly to the learner's individual learning pace. The KL value of each domain concept was determined by the results of the tests. There were two kinds of tests: (i) the tests that corresponded to each individual domain concept of the learning material (practice tests), (ii) the final tests that corresponded to the sections of the learning material (they included exercises of a variety of domain concepts). In particular, each time the learner read a domain concept, s/he had to complete a corresponding practice test. When, the learner had completed successfully all the practice tests of the domain concepts of a section (e.g. iterations with concrete number of loops, arrays, sub-programming), then s/he had to complete the final test of the section. If s/he succeeded to the final test, then s/he transited to a next section. Otherwise, s/he had advised to revise some domain concepts. Representative examples of the system's implementation follow.

- **Example 1**

George had learned the sections 1 (domain concepts 1.1 to 1.7) and 2 (domain concept 2.1) and she was taught the domain concepts of the section 3 (domain concepts 3.1 to 3.3) (Interaction I of Table 2.8). He read the concept $C_{3.1}$. Then, he was examined in the particular domain concept and succeeded 78 %. According to the above, the value of the defined membership functions for concept $C_{3.1}$ become $\mu_{Un} = 0$, $\mu_{MKn} = 0$, $\mu_{Kn} = 0.4$, $\mu_L = 0.6$ and $\mu_A = 0$. According to the FR-CN (Fig. 2.11) the concept $C_{3.1}$ affects the following concepts $C_{3.2}$ and $C_{3.3}$ with "strength of impact" 0.5 and 0.2 correspondingly. Consequently, applying the fuzzy rule R2 (b) and (c), KL($C_{3.2}$) becomes 20 % 'Known' and 30 % 'Learned'. The rest 50 % of the particular concept remains 'Unknown' (Interaction II of Table 2.8). Similarly, applying the same rules, KL($C_{3.3}$) becomes 8 % 'Known' and 12 % 'Learned'. The rest 80 % of the particular concept remains 'Unknown' (Interaction II of Table 2.8). Therefore, although concepts $C_{3.2}$ and $C_{3.3}$ are not completely unknown to George, the system advises him to read them.

- **Example 2**

Kate had learned the sections 1 (domain concepts 1.1 to 1.7), 2 (domain concept 2.1), 3 (domain concepts (3.1 to 3.3) and the concepts 4.1, 4.5 and 5.5 (Interaction I of Table 2.9). She read the concept $C_{4.2}$ to improve her knowledge level. Then, she was examined in the particular domain concept and succeeded 86 %. According to the above, the value of the defined membership functions for concept $C_{4.2}$ become $\mu_{Un} = 0$, $\mu_{MKn} = 0$, $\mu_{Kn} = 0$, $\mu_L = 0.8$ and $\mu_A = 0.2$.

Table 2.8 George's progress

Domain concepts	Learner's knowledge	
	Interaction I (μ_{Un}, μ_{MKn}, μ_{Kn}, μ_L, μ_A)	Interaction II (μ_{Un}, μ_{MKn}, μ_{Kn}, μ_L, μ_A)
1.1 Constants and variables	(0, 0, 0, 0, 1)	(0, 0, 0, 0, 1)
1.2 Assignment statement	(0,0, 0, 0.08, 0.92)	(0,0, 0, 0.08, 0.92)
1.3 Arithmetic operators	(0, 0, 0, 0, 1)	(0, 0, 0, 0, 1)
1.4 Comparative operators	(0,0, 0,0.08, 0.92)	(0,0, 0,0.08, 0.92)
1.5 Logical operators	(0, 0, 0, 0, 1)	(0, 0, 0, 0, 1)
1.6. Mathematic functions	(0, 0, 0, 0, 1)	(0, 0, 0, 0, 1)
1.7 Input-output statements	(0, 0, 0, 0, 1)	(0, 0, 0, 0, 1)
2.1 A simple program's structure	(0, 0, 0, 0, 1)	(0, 0, 0, 0, 1)
3.1 If statement	(1, 0, 0, 0, 0)	(0, 0, 0.4, 0.6, 0)
3.2 If...else if	*(1, 0, 0, 0, 0)*	*(0.5, 0, 0.2, 0.3, 0)*
3.2.1 Finding max, min	(1, 0, 0, 0, 0)	(1, 0, 0, 0, 0)
3.3 Nested if statement	*(1, 0, 0, 0, 0)*	*(0.8, 0, 0.08, 0.12, 0)*
4.1 For statement	(1, 0, 0, 0, 0)	(1, 0, 0, 0, 0)
4.2 Calc. sum in a for loop	(1, 0, 0, 0, 0)	(1, 0, 0, 0, 0)
4.3 Counting in a for loop	(1, 0, 0, 0, 0)	(1, 0, 0, 0, 0)
4.4 Calc. avrg in a for loop	(1, 0, 0, 0, 0)	(1, 0, 0, 0, 0)
4.5 Calc. max/min in a for loop	(1, 0, 0, 0, 0)	(1, 0, 0, 0, 0)
5.1 While statement	(1, 0, 0, 0, 0)	(1, 0, 0, 0, 0)
5.2 Calc. sum in a while loop	(1, 0, 0, 0, 0)	(1, 0, 0, 0, 0)
5.3 Counting in a while loop	(1, 0, 0, 0, 0)	(1, 0, 0, 0, 0)
5.4 Calc. avrg in a while loop	(1, 0, 0, 0, 0)	(1, 0, 0, 0, 0)
5.5 Calc. max/min in a while loop	(1, 0, 0, 0, 0)	(1, 0, 0, 0, 0)
5.6 Do...until	(1, 0, 0, 0, 0)	(1, 0, 0, 0, 0)
6.1 One-dimension arrays	(1, 0, 0, 0, 0)	(1, 0, 0, 0, 0)
6.2 Searching	(1, 0, 0, 0, 0)	(1, 0, 0, 0, 0)
6.3 Sorting	(1, 0, 0, 0, 0)	(1, 0, 0, 0, 0)
6.4 Two-dimensions arrays	(1, 0, 0, 0, 0)	(1, 0, 0, 0, 0)
6.5 Processing per rows	(1, 0, 0, 0, 0)	(1, 0, 0, 0, 0)
6.6 Processing per column	(1, 0, 0, 0, 0)	(1, 0, 0, 0, 0)
6.7 Processing of diagonals	(1, 0, 0, 0, 0)	(1, 0, 0, 0, 0)
7.1 Functions	(1, 0, 0, 0, 0)	(1, 0, 0, 0, 0)

According to the FR-CN (Fig. 2.11) the concept $C_{4.2}$ affects the following concepts $C_{4.3}$, $C_{4.4}$, $C_{5.2}$, $C_{5.3}$ and $C_{5.4}$ with "strength of impact" 0.45, 0.81, 1, 0.45 and 0.39 correspondingly. Consequently, applying the fuzzy rule R2 (c) and (d), KL($C_{4.3}$) becomes 36 % 'Learned' and 9 % 'Assimilated'. The rest 55 % of the particular concept remains 'Unknown' (Interaction II of Table 3.2). Similarly, applying the same rules, KL($C_{4.4}$) becomes 64.8 % 'Learned' and 16.2 % 'Assimilated' (the rest 19 % of the particular remains 'Unknown'), KL($C_{5.2}$) becomes 80 % 'Learned'

Table 2.9 Kate's progress

Domain concepts	Learner's knowledge	
	Interaction I (μ_{Un}, μ_{MKn}, μ_{Kn}, μ_L, μ_A)	Interaction II (μ_{Un}, μ_{MKn}, μ_{Kn}, μ_L, μ_A)
1.1 Constants and variables	(0, 0, 0, 0, 1)	(0, 0, 0, 0, 1)
1.2 Assignment statement	(0, 0, 0, 0, 1)	(0, 0, 0, 0, 1)
1.3 Arithmetic operators	(0, 0, 0, 0.02, 0.098)	(0, 0, 0, 0.02, 0.098)
1.4 Comparative operators	(0, 0, 0, 0, 1)	(0, 0, 0, 0, 1)
1.5 Logical operators	(0, 0, 0, 0, 1)	(0, 0, 0, 0, 1)
1.6. Mathematic functions	(0, 0, 0, 0.12, 0.88)	(0, 0, 0, 0.12, 0.88)
1.7 Input-output statements	(0, 0, 0, 0, 1)	(0, 0, 0, 0, 1)
2.1 A simple program's structure	(0, 0, 0, 0, 1)	(0, 0, 0, 0, 1)
3.1 If statement	(0, 0, 0, 0.3, 0.7)	(0, 0, 0, 0.3, 0.7)
3.2 If...else if	(0, 0, 0, 0.4, 0.6)	(0, 0, 0, 0.4, 0.6)
3.2.1 Finding max, min	(0, 0, 0, 0.1, 0.9)	(0, 0, 0, 0.1, 0.9)
3.3 Nested if statement	(0, 0, 0, 0.4, 0.6)	(0, 0, 0, 0.4, 0.6)
4.1 For statement	(0, 0, 0, 0.73, 0.27)	(0, 0, 0, 0.73, 0.27)
4.2 Calc. sum in a for loop	(1, 0, 0, 0, 0)	(0, 0, 0, 0.8, 0.2)
4.3 Counting in a for loop	*(1, 0, 0, 0, 0)*	*(0.55, 0, 0, 0.36, 0.09)*
4.4 Calc. avrg in a for loop	*(1, 0, 0, 0, 0)*	*(0.19, 0, 0, 0.648, 0.162)*
4.5 Calc. max/min in a for loop	(0, 0, 0, 0.67, 0.33)	(0, 0, 0, 0.67, 0.33)
5.1 While statement	(1, 0, 0, 0, 0)	(1, 0, 0, 0, 0)
5.2 Calc. sum in a while loop	*(1, 0, 0, 0, 0)*	*(0, 0, 0, 0.8, 0.2)*
5.3 Counting in a while loop	*(1, 0, 0, 0, 0)*	*(0.55, 0, 0, 0.36, 0.09)*
5.4 Calc. avrg in a while loop	*(1, 0, 0, 0, 0)*	*(0.61, 0, 0, 0.312, 0.078)*
5.5 Calc. max/min in a while loop	(0, 0, 0, 0.67, 0.33)	(0, 0, 0, 0.67, 0.33)
5.6 Do...until	(1, 0, 0, 0, 0)	(1, 0, 0, 0, 0)
6.1 One-dimension arrays	(1, 0, 0, 0, 0)	(1, 0, 0, 0, 0)
6.2 Searching	(1, 0, 0, 0, 0)	(1, 0, 0, 0, 0)
6.3 Sorting	(1, 0, 0, 0, 0)	(1, 0, 0, 0, 0)
6.4 Two-dimensions arrays	(1, 0, 0, 0, 0)	(1, 0, 0, 0, 0)
6.5 Processing per rows	(1, 0, 0, 0, 0)	(1, 0, 0, 0, 0)
6.6 Processing per column	(1, 0, 0, 0, 0)	(1, 0, 0, 0, 0)
6.7 Processing of diagonals	(1, 0, 0, 0, 0)	(1, 0, 0, 0, 0)
7.1 Functions	(1, 0, 0, 0, 0)	(1, 0, 0, 0, 0)

and 20 % 'Assimilated', KL($C_{5.3}$) becomes 36 % 'Learned' and 9 % 'Assimilated' (the rest 55 % of the particular concept remains 'Unknown') and KL($C_{5.4}$) becomes 31.2 % 'Learned' and 7.8 % 'Assimilated' (the rest 61 % of the particular concept remains 'Unknown') (Interaction II of Table 3.3). Therefore, the increase of Kate's knowledge level on $C_{4.2}$ improves automatically her knowledge level

Table 2.10 Nick's progress

Domain concepts	Learner's knowledge	
	Interaction I (μ_{Un}, μ_{MKn}, μ_{Kn}, μ_L, μ_A)	Interaction II (μ_{Un}, μ_{MKn}, μ_{Kn}, μ_L, μ_A)
1.1 Constants and variables	(0, 0, 0, 0, 1)	(0, 0, 0, 0, 1)
1.2 Assignment statement	(0, 0, 0, 0, 1)	(0, 0, 0, 0, 1)
1.3 Arithmetic operators	(0, 0, 0, 0.02, 0.098)	(0, 0, 0, 0.02, 0.098)
1.4 Comparative operators	(0, 0, 0, 0, 1)	(0, 0, 0, 0, 1)
1.5 Logical operators	(0, 0, 0, 0, 1)	(0, 0, 0, 0, 1)
1.6. Mathematic functions	(0, 0, 0, 0.12, 0.88)	(0, 0, 0, 0.12, 0.88)
1.7 Input-output statements	(0, 0, 0, 0, 1)	(0, 0, 0, 0, 1)
2.1 A simple program's structure	(0, 0, 0, 0, 1)	(0, 0, 0, 0, 1)
3.1 If statement	(0, 0, 0, 0.3, 0.7)	(0, 0, 0, 0.3, 0.7)
3.2 If...else if	(0, 0, 0, 0.4, 0.6)	(0, 0, 0, 0.4, 0.6)
3.2.1 Finding max, min	(0, 0, 0, 0.1, 0.9)	(0, 0, 0, 0.1, 0.9)
3.3 Nested if statement	(0, 0, 0, 0.4, 0.6)	(0, 0, 0, 0.4, 0.6)
4.1 For statement	(0, 0, 0, 0.73, 0.27)	(0, 0, 0, 0.73, 0.27)
4.2 Calc. sum in a for loop	*(0, 0, 0, 0.8, 0.2)*	*(0, 0, 1, 0, 0)*
4.3 Counting in a for loop	*(0, 0, 0, 0.6, 0.4)*	*(0, 0, 0, 1, 0)*
4.4 Calc. avrg in a for loop	*(0, 0, 0, 0.7, 0.3)*	*(0, 0, 1, 0, 0)*
4.5 Calc. max/min in a for loop	(0, 0, 0, 0.67, 0.33)	(0, 0, 0, 0.67, 0.33)
5.1 While statement	(0, 0, 0, 1, 0)	(0, 0, 0, 1, 0)
5.2 Calc. sum in a while loop	(0, 0, 0, 0.8, 0.2)	(0, 0, 1, 0, 0)
5.3 counting in a while loop	*(0, 0, 0, 0.6, 0.4)*	*(0, 0, 0.45, 0.15, 0.4)*
5.4 Calc. avrg in a while loop	*(0, 0, 0, 0.7, 0.3)*	*(0, 0, 0.81, 0, 0.19)*
5.5 Calc. max/min in a while loop	(0, 0, 0, 0.67, 0.33)	(0, 0, 0, 0.67, 0.33)
5.6 Do...until	(1, 0, 0, 0, 0)	(1, 0, 0, 0, 0)
6.1 One-dimension arrays	(1, 0, 0, 0, 0)	(1, 0, 0, 0, 0)
6.2 Searching	(1, 0, 0, 0, 0)	(1, 0, 0, 0, 0)
6.3 Sorting	(1, 0, 0, 0, 0)	(1, 0, 0, 0, 0)
6.4 Two-dimensions arrays	(1, 0, 0, 0, 0)	(1, 0, 0, 0, 0)
6.5 Processing per rows	(1, 0, 0, 0, 0)	(1, 0, 0, 0, 0)
6.6 Processing per column	(1, 0, 0, 0, 0)	(1, 0, 0, 0, 0)
6.7 Processing of diagonals	(1, 0, 0, 0, 0)	(1, 0, 0, 0, 0)
7.1 Functions	(1, 0, 0, 0, 0)	(1, 0, 0, 0, 0)

on other related domain concepts, also. Indeed, the fact that the knowledge level of concept $C_{5.2}$ became automatically from 100 % 'Unknown', 80 % 'Learned' and 20 % 'Assimilated', without Kate read it, is particular important. This change triggers the system to infer that $C_{5.2}$ is already known for Kate.

- **Example 3**

Nick had learned the sections 1 (the domain concepts 1.1 to 1.7), 2 (the domain concept 2.1), 3 (the domain concepts 3.1 to 3.3), 4 (the domain concepts 4.1 to 4.5) and some domain concepts 5.1 to 5.5 of the section 5 (Interaction I of Table 2.10). He revised the concept $C_{5.2}$. During the revision, he was examined in the particular domain concept and succeeded 73 %. According to the above, the value of the defined membership functions for concept $C_{5.2}$ become $\mu_{Un} = 0$, $\mu_{MKn} = 0$, $\mu_{Kn} = 1$, $\mu_{L} = 0$ and $\mu_{A} = 0$. According to the FR-CN (Fig. 2.11) the concept $C_{5.2}$ affects the preceding concepts $C_{4.2}$, $C_{4.3}$, $C_{4.4}$ and the following concepts $C_{5.3}$ and $C_{5.4}$ with "strength of impact" 1, 0.45, 0.81, 0.45 and 0.81 correspondingly. Consequently, applying the fuzzy rule R8 is: $x_{4.2} = (1 - 1) * 86 + \min[1 * 86, 1 * 73] = 73$. That degree of success corresponds to the fuzzy set 'Known' with $\mu_{Kn} = 1$. (Interaction II of Table 3.4). Similarly, applying the same rule, $KL(C_{4.3})$ becomes 100 % 'Learned', and $KL(C_{4.4})$ becomes 100 % 'Known' (Interaction II of Table 3.4). Furthermore, according to the rules R3 and R4 (a), $KL(C_{5.3})$ becomes 45 % 'Known', 15 % 'Learned' and 40 % 'Assimilated' and $KL(C_{5.4})$ becomes 70 % 'Known' and 30 % 'Assimilated' (Interaction II of Table 2.10).

2.5   Conclusions and Discussion

Learning is a complicated process. It cannot be accurately said that a learner knows or does not know a domain concept. For example, a new domain concept may be completely unknown to the learner but in other circumstances it may be partly known due to previous related knowledge of the learner. On the other hand, domain concepts, which were previously known by the learner, may be completely or partly forgotten. Hence, currently they may be partly known or completely unknown. In this sense, the level of knowing cannot be accurately represented. Finally, the teaching process itself changes the status of knowledge of a user. This is happened due to the fact that a learner accepts new concepts while being taught. Furthermore, the learner's knowledge is a moving target. The knowledge level of a domain concept is increased when the student's performance is improved. Alternatively, it is decreased when the student forgets. Improvement of the knowledge level of a domain concept should lead to the increase of the knowledge level of all the related concepts (prerequisite and following), with his concept. Similarly, poor performance on a domain concept should lead to decrease of the knowledge level of all the related concepts with this concept.

In view of the above, an effective adaptive tutoring system has to be responsible for tracking cognitive state transitions of learners with respect to their progress or non-progress. The alterations on the state of student's knowledge level are not linear. They deal with uncertainty. Thus, a solution to represent these is fuzzy logic. Therefore, the target of this section was to develop a rule-based fuzzy logic system, which models the cognitive state transitions of learners, such as forgetting, learning

Table 2.11 Correlation of an e-shop and an adaptive e-learning system concerning the presented rule-based fuzzy logic system

	E-shop	E-learning
Nodes	Products	Domain concepts
Arcs	Preferences' dependencies	Knowledge dependencies
Fuzzy sets	Descriptions of a preference (e.g. 'uninterested', 'interested', 'liked', 'preferred')	Descriptions of knowledge level (e.g. 'unknown', 'insufficiently known', 'known', 'learned')
Changeable states	Preferences	Knowledge level

or assimilating. The presented rule-based fuzzy logic system identifies and updates each time the student's knowledge level not only for the current concept, which is delivered to the learner, but also for all the related concepts with this concept. To achieve that, the system considers either the learner's performance or the knowledge dependencies that exist between the domain concepts of the learning material. In the particular rule-based fuzzy logic system, fuzzy sets are used in order to describe how well each individual domain concept is known and learned. Furthermore, it uses a mechanism of rules over the fuzzy sets, which is triggered after any change of the value of the knowledge level of a domain concept and updates the values of the knowledge level of all the related domain concepts with that. Therefore, the educational system, which has integrated the particular rule-based fuzzy logic system, is able to makes dynamic decisions on how the teaching syllabus is presented to the learner to fit his/her personal needs and learning pace.

The operation of the system is based on the knowledge domain representation that is implemented through a Fuzzy Related-Cognitive Network. This kind of knowledge domain representation helps to manage to represent either the order in which the domain concepts of the learning material have to be taught and organized, or the knowledge dependencies that exist between the domain concepts. This is significant because the knowledge level of a domain concept increases or decreases due to changes on the knowledge level of a related domain concept. The design of the learning material and the definition of the individual domain concepts that it includes, are based on the knowledge and experience of domain experts. Furthermore, the contribution of domain experts is significant for the definition of the knowledge dependencies that exist among the domain concepts of the learning material and their "strength o impact" on each other.

The presented rule-based fuzzy logic system is applicable to systems, in which the user's changeable state and/or preferences are affected by the existing dependencies among the system's elements (like concepts, preferences, events, choices). Thereafter, the particular system could be implemented in adaptive systems other than adaptive tutoring system. For example, it could be used in an e-shop, where the preference of an online shopper for particular products can be used in order to guess and propose her/him other products that the user is likely to be interested in. In the Table 2.11 the correlation of an e-shop and an adaptive e-learning system is presented concerning the particular rule-based fuzzy logic system (Table 2.11).

Chapter 3
A Novel Hybrid Student Model for Personalized Education

Abstract The goal of each web-based educational system is to offer effective learning such as real-classroom education and further more. To achieve this goal, the web-based educational system has to adapt dynamically to each individual student's needs and preferences. A solution to this is the student model, which allows the understanding and identification of each individual student's needs. In this chapter a novel student model, which is called F.O.S., is presented. F.O.S. combines three different student modeling approaches. It combines an overlay model with stereotypes and a rule-based mechanism. Furthermore, F.O.S. has been fully implemented in a web-based educational application, which teaches the programming language 'C'. The particular hybrid student model allows each individual learner to complete the training program in her/his own learning pace and abilities.

3.1 Introduction

The rapid development of computer and Internet technologies, and the accessibility of e-learning applications by a large and heterogeneous group of learners at any time and place, has led to a rapid and significant growth of Web-based learning environments. Web-based educational systems offer easy access to knowledge domains and learning processes from everywhere for everybody at any time. Therefore, they facilitate the access in knowledge. The goal of each web-based educational system is to offer effective learning such as real-classroom education and further more. However, traditional web-based and standalone educational systems still have several shortcomings when compared to real-life classroom teaching, such as lack of contextual and adaptive support, lack of flexible support of the presentation and feedback, lack of the collaborative support between students and systems (Xu et al. 2002). This is happened because at the classroom level, teachers can readjust each time the instructional process and the teaching strategy considering the student's needs and abilities. So, the challenge is to develop Web-based educational systems that adapt dynamically to each individual student for effective delivery of domain knowledge.

© Springer International Publishing Switzerland 2015   61
K. Chrysafiadi and M. Virvou, *Advances in Personalized Web-Based Education*,
Intelligent Systems Reference Library 78, DOI 10.1007/978-3-319-12895-5_3

Web-based educational systems offer easy access to knowledge domains and learning processes from everywhere for everybody at any time. As a result, users of web-based educational systems are of varying backgrounds. They have heterogeneous needs, different levels of knowledge and abilities. That is the reason why researches in the field of e-learning have expanded their interests on adaptive e-learning, which is suitable for teaching heterogeneous student populations (Schiaffino et al. 2008). An adaptive system must be capable of managing learning paths adapted to each user, monitoring user activities, interpreting those using specific models, inferring user needs and preferences and exploiting user and knowledge domain to dynamically facilitate the learning process (Boticario et al. 2005). In other words, an adaptive educational system has to provide personalization to the specific needs, knowledge and background of each individual student.

A solution is the student model. Student modeling has been introduced in Intelligent Tutoring Systems, but its use has been extended to most current educational software applications that aim to be adaptive and personalized. A student model allows understanding and identification of student needs. By keeping a model for every user, a system can successfully personalize its content and utilize available resources accordingly (Kyriacou 2008). For example, in an adaptive educational application, a student model can be used to achieve accurate student diagnosis and predict a student's needs. In return, it offers individualized courses (Gaudioso et al. 2010), adaptive navigation support (Castillo et al. 2009), help and feedback to students (Tsiriga and Virvou 2003a; Chrysafiadi and Virvou 2008), allowing them to learn in their own pace (Chrysafiadi and Virvou 2013c).

The student's model dimensions and properties correspond to the physical student's features and characteristics (Yang et al. 2010). Therefore, in order to construct a student model, it has to be considered what information and data about a student should be gathered. The student's characteristics are: the knowledge level, the errors and misconceptions, the learning preferences and style, other cognitive features, the emotions, the motivation and meta-cognitive skills. To model them there is a variety of student modeling techniques to choose: overlay model, stereotypes, perturbation, constraint-based model, learning machine algorithms, fuzzy logic, Bayesian networks etc. (Chrysafiadi and Virvou 2013b). However in most cases there is the need to model more than one student's characteristics. That is achieved by using a hybrid student model bringing together features of different techniques of student modeling.

3.2 Related Work

Each student modeling technique considers, usually, only one or a limited number of students' characteristics. However, a student model should consider a significant number of student's characteristics to be effective. Therefore, the need to model a variety of student's characteristics creates the need for hybrid student models. A hybrid student model allows the tutoring system to carry out the personalization efficiently. That is the reason why many adaptive and/or personalized tutoring systems perform student modeling combing different modeling techniques, like overlay model with stereotypes, stereotypes with cognitive theories, Bayesian networks with machine learning techniques etc.

Many researchers have used a hybrid student model, which brings together various features of different techniques of student modeling, in order to combine various aspects of student's characteristics. For example, Web-EasyMath (Tsiriga and Virvou 2002, 2003c) uses a combination of stereotypes with the machine learning technique of the distance weighted k-nearest neighbor algorithm, in order to initialize the model of a new student. The student is first assigned to a stereotype category concerning her/his knowledge level and then the system initializes all aspects of the student model using the distance weighted k-nearest neighbor algorithm among the students that belong to the same stereotype category with the new student. A combination of stereotypes with machine learning techniques has been, also, used in Web-PTV (Tsiriga and Virvou 2003a, b) and GIAS (Castillo et al. 2009). Furthermore, Inventado et al. (2010) and Baker et al. (2010) have used a combination of Bayesian networks and machine learning technique in order to observe students' reactions and adjust the instruction automatically to each individual learner. Also, Balakrishnan (2011) has build a student model upon ontology of machine learning strategies in order to model the effect of affect on learning and recognize for any learning task, what learning strategy, or combination thereof, is likely to be the most effective. Millán and Pérez-de-la-cruz (2002) have improved the accuracy and efficiency of the diagnosis process through a student model, which applied Bayesian networks and Adaptive Testing Theory (cognitive theory). Other adaptive and/or personalized tutoring systems that have used a compound student model, which combines Bayesian networks with cognitive theories, are: ABM (Hernández et al. 2010); AMPLIA (Viccari et al. 2008) and PlayPhysics (Muñoz et al. 2011). Virvou and Kabassi (2002) have added more "human" reasoning to F-SMILE using a novel combination of HPR (cognitive theory) with a stereotype-based mechanism. In addition, the student model of TADV (Kosba et al. 2003, 2005) combines an overlay model with fuzzy techniques, to represent the knowledge of individual students and their communication styles. Kavčič (2004a) has been used a similar combination of student modeling techniques. Furthermore, InfoMap (Lu et al. 2005, 2007) uses an overlay student model in combination with a buggy model for identification of the deficient knowledge. Also, KERMIT maintains two kinds of student models: a constraint-based model and an overlay model (Suraweera and Mitrovic 2004). Glushkova (2008) has applied a qualitative overlay student model to represent learners' knowledge level to DeLC system. However, because she wanted to model, also, learners' manner of access to training resources, their preferences, habits and behaviors during the learning process, she have combined the overlay model with stereotype modeling. A combination of stereotypes with overlay model has been performed in ELaC (Chrysafiadi and Virvou 2013c). Moreover, AUTO-COLLEAGUE (Tourtoglou and Virvou 2008, 2012) performs student modeling through a hybrid student model based on perturbation and the stereotype-based modeling technique. A combination of fuzzy logic and machine learning techniques has been used in ADAPTAPlan (Jurado et al. 2008), while overlay model has been combined with ontologies in Personal Reader (Dolog et al. 2004), OPAL (Cheung et al. 2010) and IWT (Albano 2011). It is remarkable to refer that a compound student model can include more than two student modeling techniques. For example, Surjono and Maltby (2003) have combined an overlay model with perturbation technique and stereotypes; Chrysafiadi and Virvou (2014) have combined fuzzy techniques with stereotypes and overlay model; the student model of INSPIRE (Grigoriadou et al. 2002; Papanikolaou et al. 2003) combines stereotypes and an overlay model with fuzzy logic techniques; and the student model of DEPTHS (Jeremić et al. 2012) is a combination of stereotype and overlay modeling with fuzzy rules.

Conclusions about the most common combination of student modeling techniques are drawn considering the hybrid student models of the literature review. An overlay student model usually is combined with stereotypes or fuzzy logic techniques. Stereotypes are blended, mainly, with overlay, but they are also combined with machine learning or fuzzy logic techniques. Perturbation student model is combined only with overlay and stereotypes. Machine learning techniques are used mostly to support stereotype modeling, but there is, also, an interest to combine them with Bayesian networks. Cognitive theories are applied with stereotypes and Bayesian Networks. Fuzzy logic is usually used with overlay or stereotype student models. Bayesian networks are blended, mainly, with machine learning techniques and cognitive theories, but they are, also, combined with stereotypes. Ontologies are primarily combined with overlay student modeling.

3.3 The F.O.S. Hybrid Student Model

A hybrid student model, which brings together various features of different techniques of user modeling, is the solution for offering a more adaptive learning system. The reason for this is the fact that the student model needs to combines various aspects of student's characteristics that is both domain dependent and domain independent in order to carry out the personalization efficiently (Yang et al. 2010). This way, the model not only can exhibit unique individual characteristics and preferences of each learner by monitoring and tracing the changes of their knowledge, skills, interests, but also classify the learners according to their performance, individual learning behaviors and activities (Yang et al. 2010). That is the reason for the development of a novel hybrid student, which combined overlay technique and stereotypes with fuzzy logic.

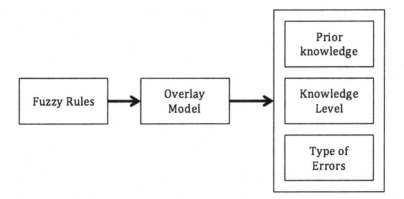

Fig. 3.1 The hybrid student model F.O.S.

The particular hybrid student model is called **F**uzzy logic system, **O**verlay and **S**tereotypes (F.O.S.). It includes a rule-based fuzzy logic system, which is responsible for tracking and updating the knowledge level of each domain concept of the learning material for each individual system; an overlay model, which represents the knowledge level of the student; and a three-dimensional stereotype model (Fig. 3.1). The fuzzy rules are used to define the learner's knowledge level of each domain concept of the learning material. Therefore, the results of the application of that rules determine the overlay model. Information of the overlay model is used to define the first dimension of the stereotype model, which concerns the learner's knowledge level and vary from novices to experts. The second dimension, which concerns the type of errors that a learner can make, and the third dimension, which concerns the prior knowledge of the student on related knowledge domain fields, are not determined by the rest parts of the student model.

3.3.1 Fuzzy Rules

It is not a straightforward task to define for each learner which concepts are unknown, known or assimilated and at what degree. The student's progress assessment includes statements like "The concept A is 72 % 'Known' for the student", "The concept X is 60 % 'insufficiently known' for the learner", "The student has learnt 100 % the concept Y", "The concept Z is 100 % 'unknown' for student Z". This information is imprecise. One possible approach to deal with this is fuzzy set techniques, with their ability to naturally represent human conceptualization. That is the reason for the integration of fuzzy logic techniques into the hybrid student model.

The rule-based fuzzy logic system is responsible for identifying and updating the student's knowledge level of all the concepts of the knowledge domain. It uses fuzzy sets to represent the student's knowledge level and a mechanism of rules over the fuzzy sets, which is triggered after a change has occurred on the student's knowledge level of a domain concept. This mechanism updates the student's knowledge level of all related with this concept, concepts. Its operation is based on the knowledge dependencies that exist between the domain concepts of the learning material and their "strength of impact" on each other.

Fuzzy sets are used to characterize the changeable user's state. For example, ("Unknown", "Known", "Learned"} or ("Unknown", "Insufficiently Known", "Known", "Learned", "Assimilated"} are the fuzzy sets of educational adaptive systems. Therefore, FS_1, FS_2, … , FS_n are the defined fuzzy sets and μ_{FS_i}, $i = 1, 2, 3, … , n$ are the corresponding membership functions. Therefore, a set $(\mu_{FS_1}, \mu_{FS_2}, \mu_{FS_3}, … , \mu_{FS_n})$ is used to express the student knowledge of a domain concept with $\mu_{FS_1} + \mu_{FS_2} + \mu_{FS_3} + \cdots + \mu_{FS_n} = 1$. The fuzzy rules are depicted in Fig. 3.2.

C_i is taught before C_j

❖ *Based on updates of the KL(C_i), the KL(C_j) is improved according to:*

   R1: If the same fuzzy sets are active for both C_i and C_j, then KL(C_j)= FS_x with $\mu_{FSx}(C_j) = \max [\mu_{FSx}(C_j), \mu_{FSx}(C_i) * \mu_D(C_i, C_j)]$, where FS_x is the last active fuzzy set. Subtract the value (new $\mu_{FSx}(C_j)$ - previous $\mu_{FSx}(C_j)$) from the others $\mu_{FSy}(C_j)$ ($FSy < FSx$) sequentially until $\Sigma \mu_{FSi} = 1$.

   R2: If KL(C_j)= FS_x and KL(C_i)=FS_y, then KL(C_j)= FS_y with $\mu_{FSy}(C_j) = \mu_{FSy}(C_i) * \mu_D(C_i, C_j)$

❖ *Based on updates of the KL(C_i), the KL(C_j) is deteriorated according to:*

   R3: If KL(C_j) = FS_n, then if $\mu_{FS1}(C_j) + \mu_{FS2}(C_j) + \cdots + \mu_{FSn-1}(C_j) < \mu_{FSi}(C_i) * \mu_D(C_i, C_j)$, where $i < n$, then the corresponding value is subtracted by $\mu_{FSn}(Cj)$, else it does not change.

   R4: If KL(C_j)= FS_y and KL(C_i)=FS_x, then KL(C_j)= FS_x with $\mu_{FSx}(C_j) = \mu_{FSx}(C_i) * \mu_D(C_i, C_j)$

❖ *Based on updates of the KL(C_j), the KL(C_i) is improved according to:*

   R5: If the same fuzzy sets are active for both C_i and C_j, then KL(C_j)= FS_x with $\mu_{FSx}(C_j) = \max [\mu_{FSx}(C_j), \mu_{FSx}(C_i) * \mu_D(C_i, C_j)]$, where FS_x is the last active fuzzy set.

   R6: If KL(C_i)= FS_x and KL(C_j)=FS_y, then KL(C_i)= FS_y with $\mu_{FSy}(C_i) = \mu_{FSy}(C_j) * \mu_D(C_j, C_i)$

❖ *Based on updates of the KL(C_j), the KL(C_i) is deteriorated according to:*

   R7: *If KL(C_i) = FS_n with $\mu_{FSn}(Ci) = 1$, then it does not change*

   R8: The formula $x_i = \left(1 - \mu_D(C_i, C_j)\right) * x_i + \min [\mu_D(C_i, C_j) * x_i, \mu_D(C_i, C_j) * x_j]$, where x_i and x_j are the values of the criterion, which determines the fuzzy sets that are active each time for C_i and C_j respectively, is used (for the calculation of previous x_i, the membership value of the upper active fuzzy set is used). Then, using the new x_i, the KL(C_i) is determined, calculating the membership functions.

❖ **Limitation:** $\Sigma \mu_{FSi} = 1$

Where FSx, FSy are fuzzy sets that represent knowledge levels with $FSx<FSy$, FS_n is the last fuzzy set (FS_1, FS_2, …, FS_n), μ_{FS} is the membership function of the fuzzy set FS, KL( ) denotes the "Knowledge Level of", $\mu_D(C_i, C_j)$ and $\mu_D(C_i, C_j)$ are the "strength of impact" of C_i on C_j and of C_i on C_j correspondingly.

Fig. 3.2 The fuzzy rules

The results of the application of the particular fuzzy rules update the overlay model of the hybrid student model.

3.3.2 Overlay Model

One of the most popular and common used student models is the overlay model. It was invented by Stansfield et al. (1976) and has been used in many systems ever since. The main assumption underlying the overlay model is that a student may have incomplete but correct knowledge of the domain. Therefore, according to the overlay modeling, the student model is a subset of the domain model (Martins et al. 2008; Vélez et al. 2008), which reflects the expert-level knowledge of the subject (Brusilovsky and Millán 2007; Liu and Wang 2007). The differences between the student's and the expert's set of knowledge are believed to be the student's lack of skills and knowledge, and the instructional objective is to eliminate these differences as much as possible (Bontcheva and Wilks 2005; Michaud and McCoy 2004; Staff 2001). Consequently, the domain is decomposed into a set of elements and the overlay model is simply a set of masteries over those elements (Nguyen and Do 2008). The pure overlay model assigns a Boolean value, yes or no, to each element, indicated whether the student knows or does not know this element, while in its modern form, an overlay model represents the degree to which the user knows such a domain element by using a qualitative measure (good-average-poor) or a quantitative measure such as the probability that the student knows the concept (Brusilovsky and Millán 2007).

A fuzzy-weighted qualitative overlay model is used in the presented hybrid student model. A qualitative weighted overlay model is an extension of the pure overlay model that can distinguish several levels of student's knowledge about each concept representing user knowledge of a concept as a qualitative value (Brusilovsky and Anderson 1998; Papanikolaou et al. 2003). In the presented novel hybrid student model, the overlay model uses qualitative values, like ('unknown', 'insufficiently known', 'known', 'learned'), which corresponds to the fuzzy sets. Furthermore, it uses a set of fuzzy values $(\mu_{FS_1}, \mu_{FS_2}, \mu_{FS_3}, \ldots, \mu_{FS_n})$, which expresses the degree in which each of the above fuzzy sets (qualitative values) are active. For example, if ('unknown', 'insufficiently known', 'known', 'learned') are the qualitative values and (0, 0.3, 0.6, 0.1) is the fuzzy values that characterize the concept C_1 in the overlay model, then it means that the particular concept is 30 % 'insufficiently known', 60 % 'known' and 10 % 'learned'. That is the reason for the name 'fuzzy-weighted qualitative overlay model'.

Figure 3.3 depicts an example of the presented fuzzy-weighted qualitative overlay model. The concepts, which are colored green, belong to the subset of the domain model that the learner knows or has assimilated. The presented fuzzy-weighted qualitative overlay model is used to model the variations of the learner's knowledge level. Particularly, it is used to inform the system which domain

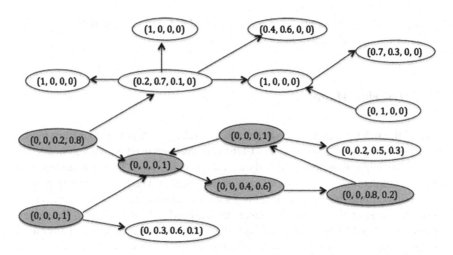

Fig. 3.3 A fuzzy-weighted qualitative overlay model [the fuzzy set in each node corresponds to the qualitative values ('unknown', 'insufficiently known', 'known', learned')]

concepts are learned, which domain concepts are partly known and which domain concepts are completely unknown.

3.3.3 Stereotypes

Another common used approach of student modeling is stereotyping. Stereotypes were introduced to user modeling by Rich (1979) in the system called GRUNDY. The main idea of stereotyping is to create groups of students with common characteristics. Such groups are called stereotypes. In other words, a stereotype normally contains the common knowledge about a group of users. A new user will be assigned into a related stereotype if some of his/her characteristics match the ones contained in the stereotype. Each stereotypes has a set of trigger conditions, which activate the stereotype if they are true, and a set of retraction conditions, which deactivate the stereotype if they are true to Kay (2000).

The stereotype student model of the presented hybrid student model is three-dimensional (Fig. 3.4). The first dimension (KL) consists of stereotypes that represent the learner's knowledge level. They vary from novices to experts. The value of KL is defined considering the information of the fuzzy-weighted qualitative overlay model. A learner is classified to a knowledge level (KL) stereotype category according to which domain concepts the learner knows and how well. The particular stereotype category gives information about the learning material that should be delivered to the learner. The second dimension (ErrTyp) consists of two stereotypes and concerns the type of errors that a learner can make. It helps the system to reason the learner's performance. For example, the system can infer if

Fig. 3.4 The three-
dimensional stereotype model

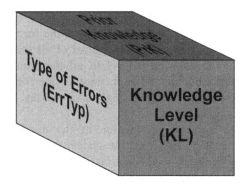

the learner reads, if s/he has difficulties in understanding, if s/he is careless, if s/he has confused with a prior knowledge on a related concept etc. Finally, the third dimension (PrK) concerns prior knowledge of the student on related knowledge domain fields. In this way, the tutoring system is able to distinguish if an error occurs due to non-learning or due to affecting by prior knowledge.

The stereotypes are updated each time new information about the learner is required. New information about the learner is obtained each time s/he interacts with the system. More concretely, each time the learner interacts with the system, s/he takes a test, the results of which determine the learner's knowledge and update her/his overlay model. The first dimension of the stereotype student model receives information from the overlay model and determines the value of KL. The stereotype categories of the second and third dimension, to which the learner should be classified, are not affected by the information that is received by the overlay model. The stereotype category of the second dimension to which the learner belongs each time, is determined by the type of errors that s/he does during the test. Also, the third dimension is determined by the learner during her/his registration to the system.

3.4 Operation of F.O.S.

When the learner interacts with the system for the first time, s/he asked to enter static information like her/his age, name and the prior knowledge (PrK) that s/he has on related fields with the knowledge domain of the system. Initially, s/he is considered to be novice. After that, the student model is updated each time new information about the learner is required. New information about the learner is obtained each time s/he interacts with the system. More concretely, each time the learner interacts with the system, s/he takes a test, the results of which determine the learner's knowledge and errors and update her/his overlay model. In particular, each time the learner interacts with the system, s/he reads a domain concept C_i and takes a test to assess her/his knowledge on the particular domain concept. The results of the test determine either the learner's knowledge on the domain concept C_i, or the type of errors (ErrTyp) that s/he made. Then, the system applies the fuzzy rules in order to identify and update the student's knowledge level of all the related concepts with C_i. The values of the knowledge level of all the domain concepts of the learning material that came off the application of the fuzzy rules are used for the definition of the fuzzy-weighted qualitative overlay model. So, the overlay student model is updated.

The new information of the overlay model triggers the KL stereotype category. Therefore, the system decides to which stereotype category of knowledge level, it should classify the learner concerning the learner's knowledge level of the domain concepts of the learning material. If s/he succeeds in the test, then s/he is transited to a next knowledge level and the value of KL is increased. Otherwise, if s/he fails, then s/he remains to the same knowledge level or s/he is returned to a previous knowledge level, according to her/his errors. In particular, if the learner's poor performance does not affect the knowledge level of other related domain concepts, which belong to a previous section, then the value of KL remains the same; otherwise the value of KL is decreased. If the learner makes errors that correspond to concepts of previous section, then the system infers that s/he has forgot something from previous sections of the learning material. In particular, if the learner makes errors, which consider concepts of previous knowledge level, then the system checks the value of the stereotype that corresponds to her/his prior knowledge on related fields. If the system decides that the errors were made due to confusion with prior knowledge, then it does not classify the learner to a previous knowledge level, but it points out the error. Otherwise, it classifies the learner to a previous knowledge level reducing the value of KL.

Therefore, the operation of F.O.S. is described in the following steps:

1. The learner registers into the system and enters her/his age, name and her/his prior knowledge (PrK) on related fields.
2. The learning material is delivered to her/him.
3. S/he is examined in domain concept C_i taking a test.
4. The system identifies the type of errors that the learner made (ErrTyp).
5. If the learner makes errors due to confusion with prior knowledge, then it does not consider them into calculating her/his performance, but it points out the error.
6. The system identifies the learner's knowledge level on the concept C_i.
7. The system applies the fuzzy rules and defined the alterations on the learner's knowledge level of all the related with C_i concepts, considering either the knowledge dependencies that exist between the domain concepts of the learning material or the current learner's knowledge level on the domain concepts of the learning material.
8. The system updated the fuzzy-weighted qualitative overlay model.
9. The system advises the overlay model and classifies the learner to the appropriate stereotype category of knowledge level (KL).

The operation of F.O.S. leads the system to make inferences about the changes of the user's state and make useful adaptation decisions, offering dynamic adaptation to users' needs. With this approach the system identifies the alterations on the state of student's knowledge level, such as forgetting or learning are represented. These states determine the progress of the learner each time. They are revealed by the transition from one stereotype of the student model to another. Thus, the system has to decide which stereotypes have to be activated and which stereotypes have to be deactivated, at each interaction of a learner with educational application.

As Tretiakov et al. (2005) have been stated, a state-chart diagram can been used to show the sequence of a student's mind. Consequently, the educational system has to construct statechart diagrams to track the cognitive state transitions

Table 3.1 The relationship between cognitive state and KL stereotype transitions

Cognitive state		Stereotype transitions
S/he does not learn	S/he doesn't read	No transition to other KL stereotype (KL remains the same)
	S/he reads but not learn	
	S/he reads but s/he has difficulty in understanding	
S/he learns		Transition to the next KL stereotype (KL increases)
S/he forgets		Transition to a previous KL stereotype (KL decreases)
S/he reaches target knowledge		Continuous transition to the advanced KL stereotypes (KL increases continuously)

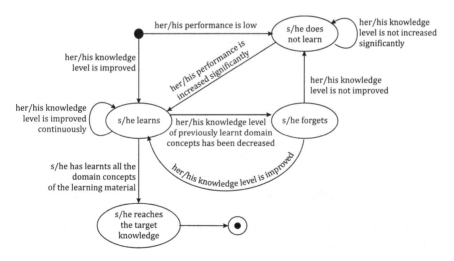

Fig. 3.5 Sequence of the changes of the learner's cognitive state in relation with her/his progress or no-progress

of learners. Table 3.1 depicts the relation between a learner's cognitive state and the transitions among the KL stereotypes of the presented hybrid student model. Furthermore, Fig. 3.5 depicts the sequence of the changes of a learner's cognitive state in relation with her/his progress or no-progress.

3.5 Application of F.O.S. in a Programming Tutoring System

The F.O.S. has been incorporated into an innovative integrated e-learning environment for computer programming and the language 'C'. The F.O.S. is responsible for identifying and updating the student's knowledge level, taking the different pace of learning of each individual learner into account. The system can adapt dynamically to each individual learner's needs by scheduling the sequence of lessons on the fly. This personalization allows each learner to complete the e-training course at their own pace and according to their ability.

In particular, the system retains static information about each student, such as her/his previous experience on computer programming and the programming languages that s/he already knows. It also retains dynamic information, such as errors, misconceptions and progress. F.O.S. allows the system to recognize when a new domain concept is completely unknown to the learner, or when it is partly known due to the learner having previous related knowledge. Furthermore, the system recognizes when a previously-known domain concept has been completely or partly forgotten by the learner. Thus it models either the possible increase or decrease of the learner's knowledge. Also, each time the system checks if the learner's errors were due to possible confusion with features of another previously-known programming language. In this case, it responds accordingly. Below, the application of F.O.S. in the programming tutoring system of the language 'C' is presented.

3.5.1 Fuzzy Rules

The defined fuzzy sets and their membership functions are the following:

- **Unknown (Un)**: the degree of success in the domain concept is from 0 to 50 %.
- **Moderate Known (MKn)**: the degree of success in the domain concept is from 40 to 70 %.
- **Known (Kn)**: the degree of success in the domain concept is from 60 to 80 %.
- **Learned (L)**: the degree of success in the domain concept is from 75 to 90 %.
- **Assimilated (A)**: the degree of success in the domain concept is from 85 to 100 %.

$$\mu_{Un} = \begin{cases} 1, & x \le 40 \\ 1 - \frac{x-40}{10}, & 40 < x < 50 \\ 0, & x \ge 50 \end{cases}$$

$$\mu_{MKn} = \begin{cases} \frac{x-40}{10}, & 40 < x < 50 \\ 1, & 50 \le x \le 60 \\ 1 - \frac{x-60}{10}, & 60 < x < 70 \\ 0, & x \le 40 \ or \ x \ge 70 \end{cases}$$

$$\mu_{Kn} = \begin{cases} \frac{x-60}{10}, & 60 < x < 70 \\ 1, & 70 \le x \le 75 \\ 1 - \frac{x-75}{5}, & 75 < x < 80 \\ 0, & x \le 60 \ or \ x \ge 80 \end{cases}$$

$$\mu_{L} = \begin{cases} \frac{x-75}{5}, & 75 < x < 80 \\ 1, & 80 \le x \le 85 \\ 1 - \frac{x-85}{5}, & 85 < x < 90 \\ 0, & x \le 75 \ or \ x \ge 90 \end{cases}$$

$$\mu_{A} = \begin{cases} \frac{x-85}{5}, & 85 < x < 90 \\ 1, & 90 \le x \le 100 \\ 0, & x \le 85 \end{cases}$$

Concerning two domain concepts C_i and C_j where C_i is taught before C_j, the knowledge level of the concepts can change according to the fuzzy rules that are depicted in Fig. 3.6 (how the knowledge level of C_j changes according to updates of the knowledge level of C_i) and Fig. 3.7 (how the knowledge level of C_i changes according to updates of the knowledge level of C_j).

3.5.2 Overlay Model

The qualitative values of the fuzzy-weighted qualitative overlay model are the defined fuzzy sets. In other words, they are the values: 'unknown', 'moderate known', 'known', 'learned' and 'assimilated'. Therefore, the overlay model uses a quintet (μ_{Un}, μ_{MKn}, μ_{Kn}, μ_L, μ_A), which expresses the degree in which each of the above qualitative values are active (Fig. 3.8). For example, (0, 0, 0.6, 0.3, 0.1) declares that the domain concept is 60 % 'known', 30 % 'learned' and 10 % 'assimilated'. Similarly, (0.7, 0.3, 0, 0, 0) declares that the concept is 70 % 'Unknown' and 30 % 'moderate known'.

Ci is taught before Cj

Based on updates of the KL(Ci), the KL(Cj) is improved according to:

R1: If the same fuzzy sets are active for both C_i and C_j, then:

  - If $KL_A(C_j) > 0$: $\mu_A(C_j) = \max[\mu_A(C_j), \mu_A(C_i) \cdot \mu_D(C_i, C_j)]$

  - Else If $KL_L(C_j) > 0$: $\mu_L(C_j) = \max[\mu_L(C_j), \mu_L(C_i) \cdot \mu_D(C_i, C_j)]$

  - Else If $KL_{Kn}(C_j) > 0$: $\mu_{Kn}(C_j) = \max[\mu_{Kn}(C_j), \mu_{Kn}(C_i) \cdot \mu_D(C_i, C_j)]$

  - Else If $KL_{MKn}(C_j) > 0$: $\mu_{MKn}(C_j) = \max[\mu_{MKn}(C_j), \mu_{MKn}(C_i) \cdot \mu_D(C_i, C_j)]$

  Subtract the value (new $\mu_x(C_j)$ – previous $\mu_x(C_j)$) from the others $\mu_y(C_j)$ sequentially until $\mu_{Un} + \mu_{MKn} + \mu_{Kn} + \mu_L + \mu_A = 1$,
  where x = {MKn, Kn, L, A} and y = {Un, MKn, Kn, L} with y < x.

R2: (a) If $KL(C_j) = Un$ and $KL(C_i) = MKn$, then $KL(C_j) = MKn$ with $\mu_{MKn}(C_j) = \mu_{MKn}(C_i) \cdot \mu_D(C_i, C_j)$

  (b) If $KL(C_j) = Un$ and $KL(C_i) = Kn$, then $KL(C_j) = Kn$ with $\mu_{Kn}(C_j) = \mu_{Kn}(C_i) \cdot \mu_D(C_i, C_j)$

  (c) If $KL(C_j) = Un$ and $KL(C_i) = L$, then $KL(C_j) = L$ with $\mu_L(C_j) = \mu_L(C_i) \cdot \mu_D(C_i, C_j)$

  (d) If $KL(C_j) = Un$ and $KL(C_i) = A$, then $KL(C_j) = A$ with $\mu_A(C_j) = \mu_A(C_i) \cdot \mu_D(C_i, C_j)$

  (e) If $KL(C_j) = MKn$ and $KL(C_i) = Kn$, then $KL(C_j) = Kn$ with $\mu_{Kn}(C_j) = \mu_{Kn}(C_i) \cdot \mu_D(C_i, C_j)$

  (f) If $KL(C_j) = MKn$ and $KL(C_i) = L$, then $KL(C_j) = L$ with $\mu_L(C_j) = \mu_L(C_i) \cdot \mu_D(C_i, C_j)$

  (g) If $KL(C_j) = MKn$ and $KL(C_i) = A$, then $KL(C_j) = A$ with $\mu_A(C_j) = \mu_A(C_i) \cdot \mu_D(C_i, C_j)$

  (h) If $KL(C_j) = Kn$ and $KL(C_i) = L$, then $KL(C_j) = L$ with $\mu_L(C_j) = \mu_L(C_i) \cdot \mu_D(C_i, C_j)$

  (i) If $KL(C_j) = Kn$ and $KL(C_i) = A$, then $KL(C_j) = A$ with $\mu_A(C_j) = \mu_A(C_i) \cdot \mu_D(C_i, C_j)$

  (j) If $KL(C_j) = L$ and $KL(C_i) = A$, then $KL(C_j) = A$ with $\mu_A(C_j) = \mu_A(C_i) \cdot \mu_D(C_i, C_j)$

Based on updates of the KL(Ci), the KL(Cj) is deteriorated according to:

R3: If $KL(C_j) = A$, then if $\mu_{Un}(C_j) + \mu_{MKn}(C_j) + \mu_{Kn}(C_j) + \mu_L(C_j) < \mu_x(C_i) \cdot \mu_D(C_i, C_j)$, where x = {Un, MKn, Kn, L}, then
the corresponding value is subtracted by $\mu_A(C_j)$, else it does not change.

R4: (a) If $KL(C_j) = L$ and $KL(C_i) = Kn$, then $KL(C_j) = Kn$ with $\mu_{Kn}(C_j) = \mu_{Kn}(C_i) \cdot \mu_D(C_i, C_j)$

  (b) If $KL(C_j) = L$ and $KL(C_i) = MKn$, then $KL(C_j) = MKn$ with $\mu_{MKn}(C_j) = \mu_{MKn}(C_i) \cdot \mu_D(C_i, C_j)$

  (c) If $KL(C_j) = L$ and $KL(C_i) = Un$, then $KL(C_j) = Un$ with $\mu_{Un}(C_j) = \mu_{Un}(C_i) \cdot \mu_D(C_i, C_j)$

  (d) If $KL(C_j) = Kn$ and $KL(C_i) = MKn$, then $KL(C_j) = MKn$ with $\mu_{MKn}(C_j) = \mu_{MKn}(C_i) \cdot \mu_D(C_i, C_j)$

  (e) If $KL(C_j) = Kn$ and $KL(C_i) = Un$, then $KL(C_j) = Un$ with $\mu_{Un}(C_j) = \mu_{Un}(C_i) \cdot \mu_D(C_i, C_j)$

  (f) If $KL(C_j) = MKn$ and $KL(C_i) = Un$, then $KL(C_j) = Un$ with $\mu_{Un}(C_j) = \mu_{Un}(C_i) \cdot \mu_D(C_i, C_j)$

Where KL( ) denotes the "Knowledge Level of", $\mu_D(C_i, C_j)$ and $\mu_D(C_i, C_j)$ are the "strength of impact" of C_i on C_j and of C_i on C_j correspondingly.

Fig. 3.6 The fuzzy rules (how the knowledge level of C_j changes according to updates of the knowledge level of C_i)

3.5.3 Stereotypes

3.5.3.1 KL Stereotype Category

The particular stereotype category is the first dimension of the third-dimension stereotype model. It concerns the knowledge level of the learner. Its value represents the expertise of the learner in the algorithms and the programming language 'C'. The value of the KL stereotype varies from "novice" users, who do not have a structural knowledge of programming and are unable to give an

---

Ci is taught before Cj

Based on updates of the KL(Cj), the KL(Ci) is improved according to:

R5: If the same fuzzy sets are active for both C_i and C_j, then:

- If $KL_A(Cj) > 0$: $\mu_A (C_j) = \max[\mu_A (C_j), \mu_A (C_i) * \mu_D(C_i, C_j)]$

- Else If $KL_L(Cj) > 0$: $\mu_L (C_j) = \max[\mu_L (C_j), \mu_L (C_i) * \mu_D(C_i, C_j)]$

- Else If $KL_{Kn}(Cj) > 0$: $\mu_{Kn} (C_j) = \max[\mu_{Kn} (C_j), \mu_{Kn} (C_i) * \mu_D(C_i, C_j)]$

- Else If $KL_{MKn}(Cj) > 0$: $\mu_{MKn} (C_j) = \max [\mu_{MKn} (C_j), \mu_{MKn} (C_i) * \mu_D(C_i, C_j)]$

 Subtract the value (new $\mu_x(C_i)$ – previous $\mu_x(C_i)$) from the others $\mu_y(C_i)$ sequentially until $\mu_{Un} + \mu_{MKn} + \mu_{Kn} + \mu_L + \mu_A = 1$, where x = {MKn, Kn, L, A} and y = {Un, MKn, Kn, L} with y < x.

R6: (a) If $KL(C_i)$ = Un and $KL(C_j)$ = MKn, then $KL(C_i)$ = MKn with $\mu_{MKn} (C_i) = \mu_{MKn} (C_j) * \mu_D(C_j, C_i)$

 (b) If $KL(C_i)$ = Un and $KL(C_j)$ = Kn, then $KL(C_i)$ = Kn with $\mu_{Kn} (C_i) = \mu_{Kn} (C_j) * \mu_D(C_j, C_i)$

 (c) If $KL(C_i)$ = Un and $KL(C_j)$ = L, then $KL(C_i)$ = L with $\mu_L(C_i) = \mu_L (C_j) * \mu_D(C_j, C_i)$

 (d) If $KL(C_i)$ = Un and $KL(C_j)$ = A, then $KL(C_i)$ = A with $\mu_A (C_i) = \mu_A (C_j) * \mu_D(C_j, C_i)$

 (e) If $KL(C_i)$ = MKn and $KL(C_j)$ = Kn, then $KL(C_i)$ = Kn with $\mu_{Kn} (C_i) = \mu_{Kn} (C_j) * \mu_D(C_j, C_i)$

 (f) If $KL(C_i)$ = MKn and $KL(C_j)$ = L, then $KL(C_i)$ = L with $\mu_L(C_i) = \mu_L (C_j) * \mu_D(C_j, C_i)$

 (g) If $KL(C_i)$ = MKn and $KL(C_j)$ = A, then $KL(C_i)$ = A with $\mu_A (C_i) = \mu_A (C_j) * \mu_D(C_j, C_i)$

 (h) If $KL(C_i)$ = Kn and $KL(C_j)$ = L, then $KL(C_i)$ = L with $\mu_L (C_i) = \mu_L (C_j) * \mu_D(C_j, C_i)$

 (i) If $KL(C_i)$ = Kn and $KL(C_j)$ = A, then $KL(C_i)$ = A with $\mu_A (C_i) = \mu_A (C_j) * \mu_D(C_j, C_i)$

 (j) If $KL(C_i)$ = L and $KL(C_j)$ = A, then $KL(C_i)$ = A with $\mu_A(C_i) = \mu_A (C_j) * \mu_D(C_j, C_i)$

Based on updates of the KL(Cj), the KL(Ci) is deteriorated according to:

R7: *If $KL(C_i)$ = A with $\mu_A(C_i)$ = 1, then it does not change.*

R8: The formula $x_i = (1 - \mu_D(C_i, C_j)) * x_i + \min [\mu_D(C_i, C_j) * x_i, \mu_D(C_i, C_j) * x_j]$, where x_i and x_j are the degree of success, which determine the fuzzy sets that are active each time for C_i and C_j respectively, is used (for the calculation of previous x_i, the membership value of the upper active fuzzy set is used). Then, using the new x_i, the $KL(C_i)$ is determined, calculating the membership functions.

Limitation **L1:** $\mu_{Un} + \mu_{MKn} + \mu_{Kn} + \mu_L + \mu_A = 1$

Where KL( ) denotes the "Knowledge Level of", $\mu_D(C_i, C_j)$ and $\mu_D(C_i, C_j)$ are the "strength of impact" of C_i on C_j and of C_i on C_j correspondingly.

Fig. 3.7 The fuzzy rules (how the knowledge level of C_i changes according to updates of the knowledge level of C_j)

acceptable answer in most cases, to "expert" users, who are able to select, use and combine programming structures creating complex programs.

For the definition of these stereotypes the conceptual framework for analyzing students' knowledge of programming that was developed by McGill and Volet (1997) and the evaluation method of knowledge of programming that was developed by deRaadt (2007) have been advised. McGill and Violet discern three knowledge types in the view of cognitive psychology: declarative (the basic knowledge of an object), procedural (how to use declarative knowledge for problem solving and decision making), strategic (upper knowledge level), and three knowledge types in the view of educational research: syntactic (basic knowledge), conceptual (be able to combine

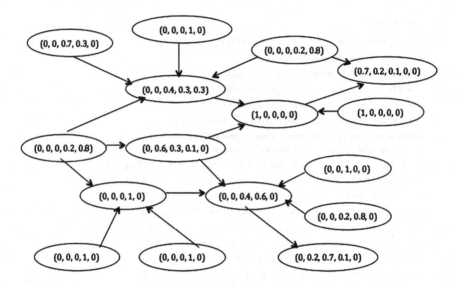

Fig. 3.8 An example of the fuzzy weighted qualitative overlay model of the presented program-ming tutoring system

knowledge, analytical thought) and strategic (integrated knowledge). De Raadt sug-gests five knowledge levels:

- **No answer**: no knowledge
- **Pre-structural**: substantial lack of knowledge
- **One-structural**: ability to describe a part of code
- **Multi-structural**: ability to describe a program line-line
- **Relational**: ability to describe the whole of a program.

Therefore, the eight stereotypes of the KL stereotype category (Table 3.2) were defined considering the above frameworks. A learner is classified to a KL stereotype category according to which chapters the learner knows and how well. That kind of information is derived from the fuzzy-weighted qualitative overlay model. The KL stereotype category to which the learner has been classified, gives information about the learning material that should be delivered to the learner (Fig. 3.9).

3.5.3.2 ErrTyp Stereotypes Category

ErrTyp stereotype takes two values: prone to syntax errors and prone to logical errors. Syntax errors are recognized if they belong in one of the following categories: anagrammatism of commands' names, omission of the definition of data, using invalid command names etc. They, usually, indicate that the learner has not read carefully and has not known adequately the chapters that correspond to her /his knowledge level. Logical errors are usually errors of design and occur

Table 3.2 Values of the KL stereotype category

KL stereotype	Knowledge type McGill and Volet (1997)	Knowledge level de Raadt (2007)
Stereotype 1: novice	No knowledge	Level 1
Stereotype 2: the learner knows the basics of the programming language C and the sequence structure of programming	Declarative—syntactic	Pre-structural
Stereotype 3: the learner knows basics of the programming language C, the sequence structure and the structures of choice	Declarative—conceptual	One-structural
Stereotype 4: the learner knows basics of the programming language C, the sequence structure, the structures of choice and the iteration structure with concrete number of loops	Procedural—syntactic	One-structural
Stereotype 5: the learner knows basics of the programming language C, the sequence structure, the structures of choice and the iteration structures with concrete or unknown number of loops	Procedural—syntactic	Multi-structural
Stereotype 6: the learner knows basics of the programming language C, the sequence structure, the structures of choice, the iteration structures and one-dimensional arrays	Procedural—conceptual	Multi-structural
Stereotype 7: the learner/he knows basics of the programming language C, the sequence structure, the structures of choice, the iteration structures and all the type of arrays	Procedural—conceptual	Relational
Stereotype 8: experts	Strategic	Relational

in case of misconceptions of the program and of the semantics and operation of the commands. They, usually, indicate that the learner has a difficulty in understanding the instructions and their logic.

3.5.3.3 PrK Stereotype Category

Furthermore, the system should known if a learner has prior knowledge of another programming language in order to be able to distinguish if an error occurs due to non-learning or due to affecting by another language. This kind of information is derived by the 3rd-dimension of the stereotype model (PrK). Therefore, the PrK stereotype category is associated with other programming languages that the learner may already know. In particular, PrK takes the following values: 'none', 'Basic', 'Pascal', 'Java'. If a learner does not know a programming language, PrK takes the value "none". If a learner knows more than one programming language of the above, PrK takes two or all these programming languages. So, it takes either the value 'none', or one or more values of the set {Basic, Pascal, Java}.

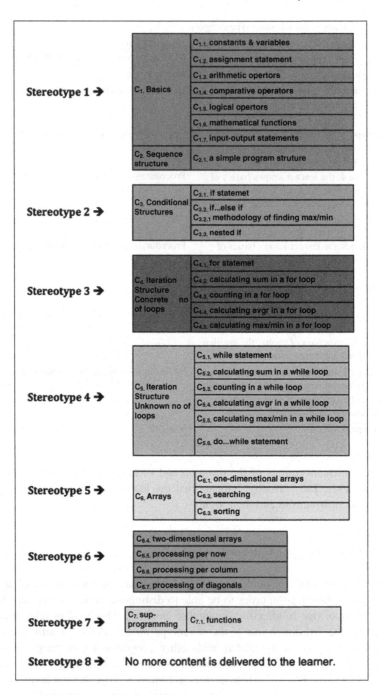

Fig. 3.9 The delivered learning material concerning the learner's KL stereotype

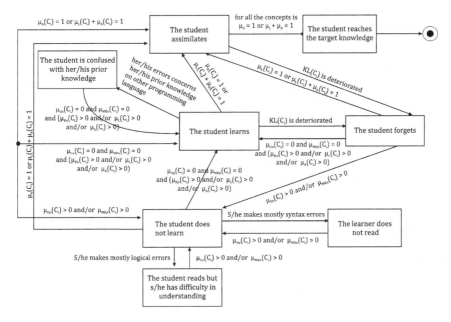

Fig. 3.10 The transitions among the learner's cognitive states

3.5.4 Cognitive State Transitions of Learners of the Programming Tutoring System

Taking into account all the above, it can be concluded that a domain concept passes through five states ('unknown', 'moderate known', 'known', 'learned', 'assimilated') during the interaction of the user with the system. Furthermore, a learner passes through several states during the learning process. S/he can learn or not, forget, assimilate or not etc. These states determine the progress of the learner each time; they determine the transition from one KL stereotype to another. Also, the transition from one KL stereotype to another can reveal the state of the learner. Figure 3.10 depicts the transitions among the learner's cognitive states for the presented programming tutoring system.

3.6   Examples of Operations

The presented programming tutoring system was used in a postgraduate program in the field of informatics at the University of Piraeus, in order to offer dynamically personalized e-training in computer programming and the language C. The learners used the system without attending any complementary course on programming, over a period of six months. They had different ages and diverse backgrounds. Examples of such backgrounds are physics, mathematics, computer science, education, human and social science.

At the beginning, all learners were considered to be novices. At the next interactions, the system delivered to them the appropriate learning material for each individual student's needs by adapting instantly to the learner's individual learning pace. The system's adaptation decisions were based on the values of the student model. The student model is updated each time the learner interacts with the system and takes a test. There were two kinds of tests: (1) the tests that corresponded to each individual domain concept of the learning material (practice tests), (2) the final tests that corresponded to the sections of the learning material (they included exercises of a variety of domain concepts). In particular, each time the learner read a domain concept (C_i), s/he had to complete a corresponding practice test. When, the learner had completed successfully ($\mu_A(C_i) = 1$ or $\mu_A(C_i) + \mu_L(C_i) = 1$) all the practice tests of the domain concepts of a section (e.g. iterations with concrete number of loops, arrays, sub-programming), then s/he had to complete the final test of the section. If s/he succeeded to the final test ($\mu_A(C_i) = 1$ or $\mu_A(C_i) + \mu_L(C_i) = 1$ for all the concept C_i of the particular section), then s/he transited to a next section. Otherwise, s/he had advised to revise some domain concepts. Representative examples of the system's implementation follow.

Example 1 Elena's current student model has the following values: KL $= 3$, ErrTyp $=$ "prone to syntax errors", PrK$=$ "none". The value KL $= 3$ comes off her current overlay model (Table 3.3, column 'before'). ErrTyp is "prone to syntax errors" due to the fact that she had made usually errors that concern anagrammatism of commands' names or invalid symbolisms of operands or commands' names. Also, PrK$=$ "none" indicates that Elena does not have previous knowledge on computer programming.

She is examining in $C_{4.2}$: "calculating sum in a 'for' loop" and is succeeding 92 %. So, the quintet, which describes Elena's knowledge level on $C_{4.2}$, is (0, 0, 0, 0, 1). However, according to the "strength of impact" of the knowledge dependencies that exist between the domain concepts of the learning material (Table 2.2), $C_{4.2}$ affects 45 % the concept $C_{4.3}$, 81 % the concept $C_{4.4}$, 100 % the concept $C_{5.2}$, 45 % the concept $C_{5.3}$, and 39 % the concept $C_{5.4}$.

According to the rule R2 (d) over the fuzzy sets (Fig. 3.6) the following occur (Table 3.3, column 'after—interaction I'):

- $\mu_A(C_{4.3}) = 0.45$ and it remains 55 % 'Unknown' ($\mu_{Un}(C_{4.3}) = 0.55$) So, the quintet for $C_{4.3}$ is (0.55, 0, 0, 0, 0.45).
- $\mu_A(C_{4.4}) = 0.81$ and it remains 19 % 'Unknown' ($\mu_{Un}(C_{4.4}) = 0.19$) So, the quintet for $C_{4.4}$ is (0.19, 0, 0, 0, 0.81).
- $\mu_A(C_{5.2}) = 1$. So, the quintet for $C_{5.4}$ is (0, 0, 0, 0, 1).
- $\mu_A(C_{5.3}) = 0.45$ and it remains 55 % 'Unknown' ($\mu_{Un}(C_{5.3}) = 0.55$) So, the quintet for $C_{5.3}$ is (0.55, 0, 0, 0, 0.45).
- $\mu_A(C_{5.4}) = 0.52$ and it remains 55 % 'Unknown' ($\mu_{Un}(C_{5.4}) = 0.48$). So, the quintet for $C_{5.4}$ is (0.48, 0, 0, 0, 0.52).

Table 3.3 Elena's progress

Domain concepts	Learner's knowledge				
	Before	After			
	$(KL = 3)$	Interaction I $(KL = 3)$	Interaction II $(KL = 3)$	Interaction III $(KL = 3)$	Interaction IV $(KL = 4)$
$C_{1.1}$	(0, 0, 0, 0, 1)	(0, 0, 0, 0, 1)	(0, 0, 0, 0, 1)	(0, 0, 0, 0, 1)	(0, 0, 0, 0, 1)
$C_{1.2}$	(0,0, 0, 0.08, 0.92)	(0,0, 0, 0.08, 0.92)	(0,0, 0, 0.08, 0.92)	(0,0, 0, 0.08, 0.92)	(0,0, 0, 0.08, 0.92)
$C_{1.3}$	(0, 0, 0, 0, 1)	(0, 0, 0, 0, 1)	(0, 0, 0, 0, 1)	(0, 0, 0, 0, 1)	(0, 0, 0, 0, 1)
$C_{1.4}$	(0,0, 0,0.08, 0.92)	(0,0, 0,0.08, 0.92)	(0,0, 0,0.08, 0.92)	(0,0, 0,0.08, 0.92)	(0,0, 0,0.08, 0.92)
$C_{1.5}$	(0, 0, 0, 0, 1)	(0, 0, 0, 0, 1)	(0, 0, 0, 0, 1)	(0, 0, 0, 0, 1)	(0, 0, 0, 0, 1)
$C_{1.6}$	(0, 0, 0, 0, 1)	(0, 0, 0, 0, 1)	(0, 0, 0, 0, 1)	(0, 0, 0, 0, 1)	(0, 0, 0, 0, 1)
$C_{1.7}$	(0, 0, 0, 0, 1)	(0, 0, 0, 0, 1)	(0, 0, 0, 0, 1)	(0, 0, 0, 0, 1)	(0, 0, 0, 0, 1)
$C_{2.1}$	(0, 0, 0, 0, 1)	(0, 0, 0, 0, 1)	(0, 0, 0, 0, 1)	(0, 0, 0, 0, 1)	(0, 0, 0, 0, 1)
$C_{3.1}$	(0, 0, 0, 0.2, 0.8)	(0, 0, 0, 0.2, 0.8)	(0, 0, 0, 0.2, 0.8)	(0, 0, 0, 0.2, 0.8)	(0, 0, 0, 0.2, 0.8)
$C_{3.2}$	(0, 0, 0, 0.6, 0.4)	(0, 0, 0, 0.6, 0.4)	(0, 0, 0, 0.6, 0.4)	(0, 0, 0, 0.6, 0.4)	(0, 0, 0, 0.6, 0.4)
$C_{3.2.1}$	(0, 0, 0, 0.3, 0.7)	(0, 0, 0, 0.3, 0.7)	(0, 0, 0, 0.3, 0.7)	(0, 0, 0, 0.3, 0.7)	(0, 0, 0, 0.21, 0.79)
$C_{3.3}$	(0, 0, 0, 0.74, 0.26)	(0, 0, 0, 0.74, 0.26)	(0, 0, 0, 0.74, 0.26)	(0, 0, 0, 0.74, 0.26)	(0, 0, 0, 0.74, 0.26)
$C_{4.1}$	(0, 0, 0, 0, 1)	(0, 0, 0, 0, 1)	(0, 0, 0, 0, 1)	(0, 0, 0, 0, 1)	(0, 0, 0, 0, 1)
$C_{4.2}$	(1, 0, 0, 0, 0)	(0, 0, 0, 0, 1)	(0, 0, 0, 0, 1)	(0, 0, 0, 0, 1)	(0, 0, 0, 0, 1)
$C_{4.3}$	(1, 0, 0, 0, 0)	(0.55, 0, 0, 0, 0.45)	(0.28, 0, 0, 0.27, 0.45)	(0, 0, 0, 0, 1)	(0, 0, 0, 0, 1)
$C_{4.4}$	(1, 0, 0, 0, 0)	(0.19, 0, 0, 0, 0.81)	(0, 0, 0, 0.6, 0.4)	(0, 0, 0, 0.55, 0.45)	(0, 0, 0, 0.55, 0.45)
$C_{4.5}$	(0.63, 0, 0, 0.11, 0.26)	(0.63, 0, 0, 0.11, 0.26)	(0.63, 0, 0, 0.11, 0.26)	(0.63, 0, 0, 0.11, 0.26)	(0, 0, 0, 0, 1)
$C_{5.1}$	(1, 0, 0, 0, 0)	(1, 0, 0, 0, 0)	(1, 0, 0, 0, 0)	(1, 0, 0, 0, 0)	(1, 0, 0, 0, 0)
$C_{5.2}$	(1, 0, 0, 0, 0)	(0, 0, 0, 0, 1)	(0, 0, 0, 0, 1)	(0, 0, 0, 0, 1)	(0, 0, 0, 0, 1)
$C_{5.3}$	(1, 0, 0, 0, 0)	(0.55, 0, 0, 0, 0.45)	(0.28, 0, 0, 0.27, 0.45)	(0, 0, 0, 0, 1)	(0, 0, 0, 0, 1)
$C_{5.4}$	(1, 0, 0, 0, 0)	(0.48, 0, 0, 0, 0.52)	(0.17, 0, 0, 0.31, 0.52)	(0.1, 0, 0, 0.18, 0.72)	(0.1, 0, 0, 0.18, 0.72)
$C_{5.5}$	(0.63, 0, 0, 0.11, 0.26)	(0.63, 0, 0, 0.11, 0.26)	(0.63, 0, 0, 0.11, 0.26)	(0.63, 0, 0, 0.11, 0.26)	(0, 0, 0, 0, 1)
$C_{5.6}$	(1, 0, 0, 0, 0)	(1, 0, 0, 0, 0)	(1, 0, 0, 0, 0)	(1, 0, 0, 0, 0)	(1, 0, 0, 0, 0)
$C_{6.1}$	(1, 0, 0, 0, 0)	(1, 0, 0, 0, 0)	(1, 0, 0, 0, 0)	(1, 0, 0, 0, 0)	(1, 0, 0, 0, 0)
$C_{6.2}$	(1, 0, 0, 0, 0)	(1, 0, 0, 0, 0)	(1, 0, 0, 0, 0)	(1, 0, 0, 0, 0)	(1, 0, 0, 0, 0)
$C_{6.3}$	(1, 0, 0, 0, 0)	(1, 0, 0, 0, 0)	(1, 0, 0, 0, 0)	(1, 0, 0, 0, 0)	(1, 0, 0, 0, 0)
$C_{6.4}$	(1, 0, 0, 0, 0)	(1, 0, 0, 0, 0)	(1, 0, 0, 0, 0)	(1, 0, 0, 0, 0)	(1, 0, 0, 0, 0)
$C_{6.5}$	(1, 0, 0, 0, 0)	(1, 0, 0, 0, 0)	(1, 0, 0, 0, 0)	(1, 0, 0, 0, 0)	(1, 0, 0, 0, 0)

(continued)

Table 3.3 (continued)

Domain concepts	Learner's knowledge				
	Before	After			
	(KL = 3)	Interaction I (KL = 3)	Interaction II (KL = 3)	Interaction III (KL = 3)	Interaction IV (KL = 4)
$C_{6.6}$	(1, 0, 0, 0, 0)	(1, 0, 0, 0, 0)	(1, 0, 0, 0, 0)	(1, 0, 0, 0, 0)	(1, 0, 0, 0, 0)
$C_{6.7}$	(1, 0, 0, 0, 0)	(1, 0, 0, 0, 0)	(1, 0, 0, 0, 0)	(1, 0, 0, 0, 0)	(1, 0, 0, 0, 0)
$C_{7.1}$	(1, 0, 0, 0, 0)	(1, 0, 0, 0, 0)	(1, 0, 0, 0, 0)	(1, 0, 0, 0, 0)	(1, 0, 0, 0, 0)

Therefore, after the interaction I Elena's student model becomes: KL = 3, ErrTyp = "prone to logical errors", PrK= "none". The value of the ErrTyp stereotype changes and becomes "prone to logical errors" due to the fact that the last errors of Elena are associated with the semantics and operation of the commands. This value indicates that Elena has a difficulty in understanding the instructions and their logic.

At the next interaction, Elena is examined in the domain concept $C_{4.4}$ and is succeeding 87 %. $C_{4.4}$ affects 100 % the concept $C_{4.2}$, 45 % the concept $C_{4.3}$, 100 % the concept $C_{5.2}$, 45 % the concept $C_{5.3}$, and 52 % the concept $C_{5.4}$. According to the fuzzy rules (Figs. 3.6 and 3.7) the following occur (Table 3.3, column 'after—interaction II'):

- According to R7, $\mu_A(C_{4.2})$ remains 1.
- According to R5 and R6 (c), $\mu_A(C_{4.3}) = 0.45$, $\mu_L(C_{4.3}) = 0.27$ and it remains 28 % 'Unknown' ($\mu_{Un}(C_{4.3}) = 0.28$). So, the quintet for $C_{4.3}$ is (0.28, 0, 0, 0.27, 0.45).
- According to R3 $\mu_A(C_{5.2})$ remains 1.
- According to R1 and R2 (c) $\mu_A(C_{5.3}) = 0.45$, $\mu_L(C_{5.3}) = 0.27$ and it remains 28 % 'Unknown' ($\mu_{Un}(C_{5.3}) = 0.28$). So, the quintet for $C_{5.3}$ is (0.28, 0, 0, 0.27, 0.45).
- According to R1 and R2 (c) $\mu_A(C_{5.4}) = 0.52$, $\mu_L(C_{5.4}) = 0.31$ and it remains 17 % 'Unknown' ($\mu_{Un}(C_{5.4}) = 0.17$). So, the quintet for $C_{5.4}$ is (0.17, 0, 0, 0.31, 0.52).

After the interaction II Elena's student model becomes: KL = 3, ErrTyp = "prone to logical errors", PrK= "none".

At the next interactions, Elena is examining in the domain concepts $C_{4.3}$ and $C_{4.5}$ and is succeeding 95 and 90 % respectively. Applying the fuzzy rules, the knowledge level of Elena changes as it is presented in Table 3.3, column 'after—interaction III and interaction IV'.

When all the domain concepts of the section 4 (i.e. the domain concepts 4.1 to 4.5) become learned, the value of the stereotype category KL of Elena's student model increases and becomes 4. Now, the domain concepts of section 5 (i.e. the domain concepts 5.1 to 5.6) should be delivered to Mary. However, the domain concepts $C_{5.2}$, $C_{5.3}$, $C_{5.4}$ and $C_{5.5}$ are considered already learned. Therefore Elena has to read only the domain concepts $C_{5.1}$ and $C_{5.6}$.

Table 3.4 Kostas' progress

Domain concepts	Learner's knowledge	
	Before	After
	(KL = 6)	(KL = 6)
$C_{1.1}$	(0, 0, 0, 0, 1)	(0, 0, 0, 0, 1)
$C_{1.2}$	(0,0, 0, 0.08, 0.92)	(0,0, 0, 0.08, 0.92)
$C_{1.3}$	(0, 0, 0, 0, 1)	(0, 0, 0, 0, 1)
$C_{1.4}$	(0,0, 0,0.08, 0.92)	(0,0, 0,0.08, 0.92)
$C_{1.5}$	(0, 0, 0, 0, 1)	(0, 0, 0, 0, 1)
$C_{1.6}$	(0, 0, 0, 0, 1)	(0, 0, 0, 0, 1)
$C_{1.7}$	(0, 0, 0, 0, 1)	(0, 0, 0, 0, 1)
$C_{2.1}$	(0, 0, 0, 0, 1)	(0, 0, 0, 0, 1)
$C_{3.1}$	(0, 0, 0, 0.2, 0.8)	(0, 0, 0, 0.2, 0.8)
$C_{3.2}$	(0, 0, 0, 0.6, 0.4)	(0, 0, 0, 0.6, 0.4)
$C_{3.2.1}$	(0, 0, 0, 0.3, 0.7)	(0, 0, 0, 0.3, 0.7)
$C_{3.3}$	(0, 0, 0, 0.74, 0.26)	(0, 0, 0, 0.74, 0.26)
$C_{4.1}$	(0, 0, 0, 0, 1)	(0, 0, 0, 0, 1)
$C_{4.2}$	(0, 0, 0, 0.12, 0.88)	(0, 0, 0, 0.12, 0.88)
$C_{4.3}$	(0, 0, 0, 0.25, 0.75)	(0, 0, 0, 0.25, 0.75)
$C_{4.4}$	(0, 0, 0, 0.24, 0.76)	(0, 0, 0, 0.24, 0.76)
$C_{4.5}$	(0, 0, 0, 0.31, 0.69)	(0, 0, 0, 0.31, 0.69)
$C_{5.1}$	(0, 0, 0, 0.3, 0.7)	(0, 0, 0, 0.3, 0.7)
$C_{5.2}$	(0, 0, 0, 0.12, 0.88)	(0, 0, 0, 0.12, 0.88)
$C_{5.3}$	(0, 0, 0, 0.25, 0.75)	(0, 0, 0, 0.25, 0.75)
$C_{5.4}$	(0, 0, 0, 0.23, 0.77)	(0, 0, 0, 0.23, 0.77)
$C_{5.5}$	(0, 0, 0, 0.31, 0.69)	(0, 0, 0, 0.31, 0.69)
$C_{5.6}$	(0, 0, 0, 0.16, 0.84)	(0, 0, 0, 0.16, 0.84)
$C_{6.1}$	(0, 0, 0.15, 0.6, 0.25)	(0, 0, 0.15, 0.6, 0.25)
$C_{6.2}$	(0, 0, 0.5, 0.5, 0)	(0, 0, 0.5, 0.5, 0)
$C_{6.3}$	(0, 0, 0.3, 0.43, 0.27)	(0, 0, 0.3, 0.43, 0.27)
$C_{6.4}$	(0, 0, 0.2, 0.3, 0.5)	(0, 0, 0, 0.5, 0.5)
$C_{6.5}$	(0, 0, 0.73, 0.1, 0.17)	(0, 0, 0, 0.8, 0.2)
$C_{6.6}$	(0, 0, 0.73, 0.1, 0.17)	(0, 0, 0.11, 0.63, 0.17)
$C_{6.7}$	(1, 0, 0, 0, 0)	(1, 0, 0, 0, 0)
$C_{7.1}$	(1, 0, 0, 0, 0)	(1, 0, 0, 0, 0)

Example 2 Kostas' current student model has the following values: KL = 6, ErrTyp = "prone to logical errors", PrK= "{Basic, Pascal}". The value KL = 6 comes off his current overlay model (Table 3.4, column 'before'). ErrTyp is "prone to logical errors" due to the fact that he had made usually errors that

concern the semantics and operation of the commands. Also, PrK$=$ "{Basic, Pascal}" indicates that Kostas knows the programming languages 'Basic' and 'Pascal', already.

He is examining in $C_{6.5}$: "processing arrays per row" and is succeeding 86 %. So, the quintet, which describes Kostas' knowledge level on $C_{6.5}$, is (0, 0, 0, 0.8, 0.2). However, according to the "strength of impact" of the knowledge dependencies that exist between the domain concepts of the learning material (Table 2.2), $C_{6.5}$ affects 99 % the concept $C_{6.4}$ and 77 % the concept $C_{6.6}$. According to the fuzzy rules (Figs. 3.6 and 3.7) the following occur (Table 3.4, column 'after'):

- According to R5 and R6 (h) are $\mu_L(C_{6.4}) = 0.79$ and $\mu_A(C_{6.4}) = 0.5$. However, due to limitation L1 is $\mu_L(C_{6.4}) = 0.5$. So, the quintet for $C_{6.4}$ is (0, 0, 0, 0.5, 0.5).
- According to R1 and R2 (h) are $\mu_L(C_{6.6}) = 0.62$, $\mu_A(C_{6.4}) = 0.17$ and it remains 11 % 'Known' ($\mu_{Kn}(C_{6.6}) = 0.11$). So, the quintet for $C_{6.6}$ is (0, 0, 0.11, 0.62, 0.17).

Consequently, Kostas remains to the same knowledge level (KL $= 6$), but the system infers that he does not need to read the domain concepts $C_{6.4}$.

Example 3 Stella's current student model has the following values: KL $= 3$, ErrTyp $=$ "prone to logical errors", PrK$=$ "none". The value KL $= 3$ comes off her current overlay model (Table 3.5, column 'before'). ErrTyp is "prone to logical errors" due to the fact that she had made usually errors that concern the semantics and operation of the commands. PrK$=$ "none" indicates that Stella does not have previous knowledge on computer programming.

She is examining in $C_{4.2}$: "calculating sum in a 'for' loop" and is succeeding 95 %. So, the quintet, which describes Stella's knowledge level on $C_{4.2}$, is (0, 0, 0, 0, 1). However, according to the "strength of impact" of the knowledge dependencies that exist between the domain concepts of the learning material (Table 2.2), $C_{4.2}$ affects 45 % the concept $C_{4.3}$, 81 % the concept $C_{4.4}$, 100 % the concept $C_{5.2}$, 45 % the concept $C_{5.3}$, and 39 % the concept $C_{5.4}$.

According to the fuzzy rules (Figs. 3.6 and 3.7) the following occur (Table 3.5, column 'after'):

- According to R2 (j) is are $\mu_A(C_{4.3}) = 0.45$. However, the current value of $\mu_A(C_{4.3})$ is 0.75. Therefore, according to R1 $\mu_A(C_{4.3}) = 0.75$ and $\mu_L(C_{4.3}) = 0.25$ So, the quintet for $C_{4.3}$ is (0, 0, 0, 0.25, 0.75).
- According to R2 (j) IS $\mu_A(C_{4.4}) = 0.81$ and it remains 19 % 'Learned' ($\mu_L(C_{4.4}) = 0.19$). So, the quintet for $C_{4.4}$ is (0, 0, 0, 0.19, 0.81).
- According to R2 (j) is $\mu_A(C_{5.2}) = 1$. So, the quintet for $C_{5.2}$ is (0, 0, 0, 0, 1).
- According to R2 (j) $\mu_A(C_{5.3}) = 0.45$ and it remains 35 % 'Learned' ($\mu_L(C_{5.3}) = 0.35$) So, the quintet for $C_{5.3}$ is (0, 0, 0, 0.35, 0.45).
- According to R2 (j) $\mu_A(C_{5.4}) = 0.39$ and it remains 61 % 'Learned' ($\mu_{Un}(C_{5.4}) = 0.61$). So, the quintet for $C_{5.4}$ is (0, 0, 0, 0.61, 0.39).

Table 3.5 Stella's progress

Domain concepts	Learner's knowledge	
	Before	After
	(KL = 3)	(KL = 5)
$C_{1.1}$	(0, 0, 0, 0.1, 0.9)	(0, 0, 0, 0.1, 0.9)
$C_{1.2}$	(0,0, 0, 0.32, 0.68)	(0,0, 0, 0.32, 0.68)
$C_{1.3}$	(0, 0, 0, 0.12, 0.88)	(0, 0, 0, 0.12, 0.88)
$C_{1.4}$	(0,0, 0,0.02, 0.98)	(0,0, 0,0.02, 0.98)
$C_{1.5}$	(0, 0, 0, 0.23, 0.77)	(0, 0, 0, 0.23, 0.77)
$C_{1.6}$	(0, 0, 0, 0.3, 0.7)	(0, 0, 0, 0.3, 0.7)
$C_{1.7}$	(0, 0, 0, 0, 1)	(0, 0, 0, 0, 1)
$C_{2.1}$	(0, 0, 0, 0.16, 0.84)	(0, 0, 0, 0.16, 0.84)
$C_{3.1}$	(0, 0, 0, 0.2, 0.8)	(0, 0, 0, 0.2, 0.8)
$C_{3.2}$	(0, 0, 0, 0.6, 0.4)	(0, 0, 0, 0.6, 0.4)
$C_{3.2.1}$	(0, 0, 0, 0, 1)	(0, 0, 0, 0, 1)
$C_{3.3}$	(0, 0, 0, 0.42, 0.58)	(0, 0, 0, 0.42, 0.58)
$C_{4.1}$	(0, 0, 0, 0.18, 82)	(0, 0, 0, 0.18, 82)
$C_{4.2}$	(0, 0, 0, 1, 0)	(0, 0, 0, 0, 1)
$C_{4.3}$	(0, 0, 0, 0.25, 0.75)	(0, 0, 0, 0.25, 0.75)
$C_{4.4}$	(0, 0, 0, 1, 0)	(0, 0, 0, 0.19, 0.81)
$C_{4.5}$	(0, 0, 0, 0, 1)	(0, 0, 0, 0, 1)
$C_{5.1}$	(0, 0, 0, 0.3, 0.7)	(0, 0, 0, 0.3, 0.7)
$C_{5.2}$	(0, 0, 0, 1, 0)	(0, 0, 0, 0, 1)
$C_{5.3}$	(0, 0, 0, 0.45, 0.35)	(0, 0, 0, 0.35, 0.45)
$C_{5.4}$	(0, 0, 0, 0.81, 0.19)	(0, 0, 0, 0.61, 0.39)
$C_{5.5}$	(0, 0, 0, 0, 1)	(0, 0, 0, 0, 1)
$C_{5.6}$	(0, 0, 0, 0, 1)	(0, 0, 0, 0, 1)
$C_{6.1}$	(1, 0, 0, 0, 0)	(1, 0, 0, 0, 0)
$C_{6.2}$	(1, 0, 0, 0, 0)	(1, 0, 0, 0, 0)
$C_{6.3}$	(1, 0, 0, 0, 0)	(1, 0, 0, 0, 0)
$C_{6.4}$	(1, 0, 0, 0, 0)	(1, 0, 0, 0, 0)
$C_{6.5}$	(1, 0, 0, 0, 0)	(1, 0, 0, 0, 0)
$C_{6.6}$	(1, 0, 0, 0, 0)	(1, 0, 0, 0, 0)
$C_{6.7}$	(1, 0, 0, 0, 0)	(1, 0, 0, 0, 0)
$C_{7.1}$	(1, 0, 0, 0, 0)	(1, 0, 0, 0, 0)

Consequently, the system infers that Stella's knowledge level has increased in more concepts than $C_{4.2}$. Thereby, it advises her overlay model and increases the value of KL stereotype, classifying her to an upper knowledge level to read the following learning material. In particular, the value of KL becomes 5 declaring that the system considers that Stella does not need to read concepts that correspond to knowledge level 4.

Example 4 The parameters of Dimitris' student model had the following values: KL = 6, ErrTyp = "prone to syntax errors", PrK= "Pascal". After having completed a test which involved exercises on two-dimensional arrays, the system discovered that he made more than 40 % errors on the assignment statement, and more specifically he used the symbol := rather than the symbol =. The system checked the values of the parameter PrK and was informed that he already knows the programming language Pascal. Thus, it was assumed that Dimitris used the symbol := for assignment due to his previous knowledge on Pascal. So, the system' reaction was to stress the error, but it did not classify him to the knowledge level 1.

Example 5 Alexis's current student model has the following values: KL = 4, ErrTyp = "prone to logical errors", PrK= "Pascal". The value KL = 4 comes off his current overlay model (Table 3.6, column 'before'). ErrTyp is "prone to logical errors" due to the fact that he had made usually errors that concern the semantics and operation of the commands. PrK= "none" indicates that Alexis knows the programming language 'Pascal'.

He is examining in $C_{5.4}$: "calculating average in while 'for' loop" and is succeeding 76 %. So, the quintet, which describes his knowledge level on $C_{5.4}$, is (0, 0, 0.8, 0.2, 0). However, according to the "strength of impact" of the knowledge dependencies that exist between the domain concepts of the learning material (Table 2.2), $C_{5.4}$ affects the concept $C_{4.2}$, $C_{4.3}$, $C_{4.4}$, $C_{5.2}$ and $C_{5.3}$.

According to the fuzzy rules (Figs. 3.6 and 3.7) the knowledge level of the related concepts is deteriorated as follows (Table 3.6, column 'after'):

- According to R7 $C_{4.2}$ remains 100 % 'Assimilated'.
- According to R8 is $x_{4.3} = (1 - 0.41) \times 88.6 + \min[0.41 \times 88.6, 0.41 \times 76] = 83.46$. Therefore, $KL(C_{4.3}) = L$ with $\mu_L(C_{4.3}) = 1$. So, the quintet for $C_{4.3}$ is $(0, 0, 0, 1, 0)$.
- According to R8 is $x_{4.4} = (1 - 0.52) \times 89 + \min[0.52 \times 89, 0.52 \times 76] = 84.8$. Therefore, $KL(C_{4.4}) = L$ with $\mu_L(C_{4.4}) = 1$. So, the quintet for $C_{4.3}$ is $(0, 0, 0, 1, 0)$.
- According to R7 $C_{5.2}$ remains 100 % 'Assimilated'.
- According to R8 is $x_{5.3} = (1 - 0.41) \times 88.6 + \min[0.41 \times 88.6, 0.41 \times 76] = 83.44$. Therefore, $KL(C_{5.3}) = L$ with $\mu_L(C_{5.3}) = 1$. So, the quintet for $C_{4.3}$ is $(0, 0, 0, 1, 0)$.

Thereby, the system infers that Alexis has forgotten the concepts $C_{4.3}$, $C_{4.4}$, $C_{5.2}$ and $C_{5.3}$ and it classifies him to the previous corresponding knowledge level to revise them, reducing the value of KL to 3.

Example 6 George, a novice student, interacted with the system for the first time. He completed the test that involved exercises, which concerned the basic concepts of "C". The results are depicted in Fig. 3.11. Particularly, he does 29 % errors

Table 3.6 Alexis' progress

Domain Concepts	Learner's Knowledge	
	Before	After
	(KL = 4)	(KL = 3)
$C_{1.1}$	(0, 0, 0, 0, 1)	(0, 0, 0, 0, 1)
$C_{1.2}$	(0,0, 0, 0.4, 0.6)	(0,0, 0, 0.4, 0.6)
$C_{1.3}$	(0, 0, 0, 0. 2, 0.8)	(0, 0, 0, 0. 2, 0.8)
$C_{1.4}$	(0,0, 0,0.15, 0.85)	(0,0, 0,0.15, 0.85)
$C_{1.5}$	(0, 0, 0, 0, 1)	(0, 0, 0, 0, 1)
$C_{1.6}$	(0, 0, 0, 0, 1)	(0, 0, 0, 0, 1)
$C_{1.7}$	(0, 0, 0, 0.43, 0.57)	(0, 0, 0, 0.43, 0.57)
$C_{2.1}$	(0, 0, 0, 0.12, 0.88)	(0, 0, 0, 0.12, 0.88)
$C_{3.1}$	(0, 0, 0, 0.2, 0.8)	(0, 0, 0, 0.2, 0.8)
$C_{3.2}$	(0, 0, 0, 0.4, 0.6)	(0, 0, 0, 0.4, 0.6)
$C_{3.2.1}$	(0, 0, 0, 0.3, 0.7)	(0, 0, 0, 0.3, 0.7)
$C_{3.3}$	(0, 0, 0, 0.52, 0.48)	(0, 0, 0, 0.52, 0.48)
$C_{4.1}$	(0, 0, 0, 0.3, 0.7)	(0, 0, 0, 0.3, 0.7)
$C_{4.2}$	(0, 0, 0, 0, 1)	(0, 0, 0, 0, 1)
$C_{4.3}$	(0, 0, 0, 0.28, 0.72)	(0, 0, 0, 1, 0)
$C_{4.4}$	(0, 0, 0, 0.2, 0.8)	(0, 0, 0, 1, 0)
$C_{4.5}$	(0, 0, 0, 0.1, 0.9)	(0, 0, 0, 0.1, 0.9)
$C_{5.1}$	(0, 0, 0, 0.3, 0.7)	(0, 0, 0, 0.3, 0.7)
$C_{5.2}$	(0, 0, 0, 0, 1)	(0, 0, 0, 0, 1)
$C_{5.3}$	(0, 0, 0, 0.28, 0.72)	(0, 0, 0, 1, 0)
$C_{5.4}$	(0, 0, 0.4, 0.2, 0.4)	(0, 0, 0.8, 0.2, 0)
$C_{5.5}$	(0, 0, 0, 0.1, 0.9)	(0, 0, 0, 0.1, 0.9)
$C_{5.6}$	(0, 0, 0, 0.47, 0.53)	(0, 0, 0, 0.47, 0.53)
$C_{6.1}$	(1, 0, 0, 0, 0)	(1, 0, 0, 0, 0)
$C_{6.2}$	(1, 0, 0, 0, 0)	(1, 0, 0, 0, 0)
$C_{6.3}$	(1, 0, 0, 0, 0)	(1, 0, 0, 0, 0)
$C_{6.4}$	(1, 0, 0, 0, 0)	(1, 0, 0, 0, 0)
$C_{6.5}$	(1, 0, 0, 0, 0)	(1, 0, 0, 0, 0)
$C_{6.6}$	(1, 0, 0, 0, 0)	(1, 0, 0, 0, 0)
$C_{6.7}$	(1, 0, 0, 0, 0)	(1, 0, 0, 0, 0)
$C_{7.1}$	(1, 0, 0, 0, 0)	(1, 0, 0, 0, 0)

Variables & Constants	A simple program's structure	Input-Output statements	Assignment statement	Arithmetic Operators & Precedence	Mathematic Functions	Comparative Operators	Logical Operators	TOTAL SYNTAX ERROS	TOTAL LOGICAL ERRORS
29%	0%	0%	33%	50%	0%	0%	20%	71%	29%
SYNTAX 100%	SYNTAX 0%	SYNTAX 0%	SYNTAX 50%	SYNTAX 100%	SYNTAX 0%	SYNTAX 0%	SYNTAX 0%		
LOGICAL 0%	LOGICAL 0%	LOGICAL 0%	LOGICAL 50%	LOGICAL 0%	LOGICAL 0%	LOGICAL 0%	LOGICAL 100%		

Fig. 3.11 Test's results of George

on variables and constants, 33 % errors on the assignment statement and 50 % errors on the arithmetic operators. Therefore, his current student model has the following values: KL = 1, ErrTyp = "prone to syntax errors", PrK= "none". The system infers that George did not read the corresponding domains. So, he is not transmitted to the next KL stereotype, but he is advised to revise the concepts of his current KL stereotype.

Table 3.7 Anna's progress

Domain concepts	Learner's knowledge	
	Before	After
	(KL = 2)	(KL = 2)
$C_{1.1}$	(0, 0, 0, 0, 1)	(0, 0, 0, 0, 1)
$C_{1.2}$	(0,0, 0, 0.2, 0.8)	(0,0, 0, 0.2, 0.8)
$C_{1.3}$	(0, 0, 0, 0.1, 0.9)	(0, 0, 0, 0.1, 0.9)
$C_{1.4}$	(0,0, 0, 0, 1)	(0,0, 0, 0, 1)
$C_{1.5}$	(0, 0, 0, 0.08, 0.82)	(0, 0, 0, 0.08, 0.82)
$C_{1.6}$	(0, 0, 0, 0.3, 0.7)	(0, 0, 0, 0.3, 0.7)
$C_{1.7}$	(0, 0, 0, 0.23, 77)	(0, 0, 0, 0.23, 77)
$C_{2.1}$	(0, 0, 0, 0.16, 0.84)	(0, 0, 0, 0.16, 0.84)
$C_{3.1}$	(0, 0, 0, 0.62, 0.38)	(0, 0, 0.11, 0.89, 0)
$C_{3.2}$	(0, 0, 0, 1, 0)	(0, 0, 1, 0, 0)
$C_{3.2.1}$	(0, 0, 0, 0.2, 0.8)	(0, 0, 0, 0.2, 0.8)
$C_{3.3}$	(0, 0, 0.3, 0.7, 0)	(0, 0, 0.64, 0.36, 0)
$C_{4.1}$	(1, 0, 0, 0, 0)	(1, 0, 0, 0, 0)
$C_{4.2}$	(1, 0, 0, 0, 0)	(1, 0, 0, 0, 0)
$C_{4.3}$	(1, 0, 0, 0, 0)	(1, 0, 0, 0, 0)
$C_{4.4}$	(1, 0, 0, 0, 0)	(1, 0, 0, 0, 0)
$C_{4.5}$	(0.63, 0, 0, 0.07, 0.3)	(0.63, 0, 0, 0.07, 0.3)
$C_{5.1}$	(1, 0, 0, 0, 0)	(1, 0, 0, 0, 0)
$C_{5.2}$	(1, 0, 0, 0, 0)	(1, 0, 0, 0, 0)
$C_{5.3}$	(1, 0, 0, 0, 0)	(1, 0, 0, 0, 0)
$C_{5.4}$	(1, 0, 0, 0, 0)	(1, 0, 0, 0, 0)
$C_{5.5}$	(0.63, 0, 0, 0.07, 0.3)	(0.63, 0, 0, 0.07, 0.3)
$C_{5.6}$	(1, 0, 0, 0, 0)	(1, 0, 0, 0, 0)
$C_{6.1}$	(1, 0, 0, 0, 0)	(1, 0, 0, 0, 0)
$C_{6.2}$	(1, 0, 0, 0, 0)	(1, 0, 0, 0, 0)
$C_{6.3}$	(1, 0, 0, 0, 0)	(1, 0, 0, 0, 0)
$C_{6.4}$	(1, 0, 0, 0, 0)	(1, 0, 0, 0, 0)
$C_{6.5}$	(1, 0, 0, 0, 0)	(1, 0, 0, 0, 0)
$C_{6.6}$	(1, 0, 0, 0, 0)	(1, 0, 0, 0, 0)
$C_{6.7}$	(1, 0, 0, 0, 0)	(1, 0, 0, 0, 0)
$C_{7.1}$	(1, 0, 0, 0, 0)	(1, 0, 0, 0, 0)

Example 7 Anna's current student model has the following values: KL $= 2$, ErrTyp $=$ "prone to logical errors", PrK$=$ "Pascal". The value KL $= 2$ comes off her current overlay model (Table 3.7, column 'before'). ErrTyp is "prone to logical errors" due to the fact that she had made usually errors that concern the semantics and operation of the commands. PrK$=$ "none" indicates that Anna knows the programming language 'Pascal'.

She is examining in $C_{3.2}$: "if...else if" and is succeeding 72 %. So, the quintet, which describes Anna's knowledge level on $C_{3.2}$, is $(0, 0, 1, 0, 0)$. However, according to the "strength of impact" of the knowledge dependencies that exist between the domain concepts of the learning material (Table 2.2), $C_{3.2}$ affects 100 % the concept $C_{3.1}$ ($C_{3.1}$ affects 50 % $C_{3.2}$) and 64 % the concept $C_{3.3}$.

According to the fuzzy rules (Figs. 3.6 and 3.7) the following occur (Table 3.7, column 'after'):

- According to R8 is $x_{3.1} = (1 - 0.5) \times 86.9 + \min[0.5 \times 86.9, 0.5 \times 72] = 79.45$. Therefore, $\mu_{Kn}(C_{3.1}) = 0.11$ and $\mu_L(C_{3.1}) = 0.89$. So, the quintet for $C_{4.3}$ is $(0, 0, 0.11, 0.89, 0)$.
- According to R4 (a) $\mu_{Kn}(C_{5.4}) = 0.64$ and it remains 36 % 'Learned' ($\mu_L(C_{5.4}) = 0.36$). So, the quintet for $C_{5.4}$ is $(0, 0, 0.64, 0.36, 0)$.

Consequently, Anna's knowledge level has been deteriorated. Therefore, the system does not transit Anna to another knowledge level stereotype category (KL remains 2). It consults her to revise the above domain concepts. Furthermore, Anna made errors concerning the equality operator. In particular, she used the symbol "$=$" rather that "$==$". However, the system does not consult her to revise the corresponding domain concept. It informs her only about the particular error. This is happened, due to the fact the value of PrK of Anna's student model is 'Pascal'. Therefore, the system infers that Anna used the symbol "$=$" for equality operator due to confusion with her previous knowledge on Pascal.

3.7 Conclusions

In this section, a novel hybrid student model was presented. The presented student model combines an overlay model and stereotypes with fuzzy logic techniques. In particular, the student model is based on an overlay model, which represents the knowledge level of a learner. The determination of the student's knowledge level of each domain concept, as well as the updating of the student model are based on the fuzzy logic technique that have been incorporated into the student model. Fuzzy sets are used in order to describe how well each individual domain concept is known, learned and assimilated. In addition, the student model includes a mechanism of rules over the fuzzy sets, which is triggered after any change of the value of the knowledge level of a domain concept and updates the values of the knowledge level of all the related domain concepts with this.

According to the learner's knowledge level and errors, the system attached her/him to the appropriate stereotype category, which concerns the student's knowledge level. The transition of a learner from one stereotype category of knowledge level to another depicts the state of the learner. In other words, the transition of a learner from one stereotype category of knowledge level to another reveals if s/he has learned or not a domain concept, if s/he has forgot a concept or if s/he has assimilated it. In addition, the presented model includes two more stereotype categories. The one category concerns the type of errors and helps the system to reason the poor performance of the learner. The other stereotype category concerns the student's prior knowledge on other related knowledge domains and helps the system either to identify the learner's knowledge level or to infer if the student's errors are caused by confusion with her/his prior knowledge.

The presented novel student model approach has been fully implemented in a web-based educational application, which teaches the programming language 'C'. The presented system provides adaptation of the instructional material, taking into account the individuality of learners in terms of background, skills and pace of learning. The particular student model includes a fuzzy-weighted qualitative overlay model; a rule-based fuzzy system; and a three-dimensional stereotype model. The first dimension of the three-dimensional stereotype model consists of eight stereotypes that represent the learner's knowledge level; the second dimension consists of two stereotypes that concern the type of programming errors (logical or syntactic) and the third dimension concerns prior knowledge of the student on other programming languages.

The particular student model allows each learner to complete the e-learning course at their own pace, taking decisions about which concepts should be delivered, which concepts need revision and which concepts are known and do not need rereading. In this way, the system helps learners to save time and effort during the learning process. The system identifies the alterations on the state of students' knowledge level, recognizes the misconceptions and needs of a learner, and reasons them. It tracks the cognitive state transitions of learners by constructing automatically state-chart diagrams. Thereby, the system recognizes if a student learns or not, if s/he reads or not, if s/he has difficulty in understanding, if s/he forgets, if s/he has confused with other programming languages that s/he has previously learned.

Chapter 4
Evaluation

Abstract In this chapter the evaluation of the hybrid student model F.O.S., which incorporates fuzzy logic techniques, is presented. The presented evaluation process is performed applying the evaluation framework PERSIVA (Chrysafiadi and Virvou 2013a), which includes both questionnaires and observations through experiments. In particular, either the educational impact (i.e. performance, satisfaction, change of learners' attitudes) or the effectiveness and validity adaptation of the personalized and/or adaptive tutoring system are assessed. The evaluated hybrid student model was integrated in a programming tutoring system for the programming language 'C'. Students of a postgraduate program in the field of Informatics on the University of Piraeus, Greece, used the particular system to learn how to program with the programming language 'C'. The evaluation results were very encouraging. They demonstrated that the presented student modeling approach had a positive impact on the learners' performance and on the learning process. Furthermore, they showed that the system made valid and meaningful adaptation decisions.

4.1 Introduction

The past decades the interest on web-based learning environments and tools has been witnessed a rapid growth. The lack of time and place limitations and the ability of web-based educational systems to offer instructions to large and heterogeneous groups of learners play a major role in the rapid growth of this interest. However, web-based educational systems and computer tutors have to react and make decisions like human tutors, in order to offer as effective learning as a real-life classroom educational process. A solution to this is the technology of Intelligent Tutoring Systems (ITSs), which are computer-based tutors that aim to provide the same level of student specific help as a human tutor (Mitrovic et al. 2007), making learning process more adaptive and student oriented (Jarusek and Pelánek 2012). A core component in any intelligent or adaptive tutoring system that represents many of the student's features, such as knowledge and individual traits, is the

© Springer International Publishing Switzerland 2015   91
K. Chrysafiadi and M. Virvou, *Advances in Personalized Web-Based Education*,
Intelligent Systems Reference Library 78, DOI 10.1007/978-3-319-12895-5_4

student model (Brusilovsky and Millán 2007). It affects automated tutoring systems in making instructional decisions (Li et al. 2011), since a student model ena-
bles understanding and identification of students' needs (Sucar and Noguez 2008).

Although, the adaptation generated by student modeling techniques often tend
to improve the interaction of the learner with the educational system, most of the
time the exploitation of such techniques makes the system more complex, less
predictable and buggier. As a consequence, it should be evaluated whether or not
the student model really improves the system (Gena 2005; Chin 2001). Therefore,
the evaluation of a student model is a crucial factor. Even though the evaluation
of adaptive systems is a difficult task due to the complexity of such systems, as
shown by many studies (Lavie et al. 2005; Markham et al. 2003; Del Missier and
Ricci 2003), several researchers have attempted to assess the student model of
their adaptive system. An assessment of the student model that SQL-Tutor uses is
presented in Mitrovic et al. (2002). Also, Weibelzahl and Weber (2003) performed
the evaluation of the accuracy of the student model of an adaptive learning system,
called the HTML-Tutor. A more recent attempt to assess the effectiveness and the
accuracy of the student model, which was applied in an ITS for learning software
design patterns, was done by Jeremić et al. (2009).

Although, there are many evaluation methods available in literature review,
as Mulwa et al. (2011) have mentioned, there is no a standard agreed measurement framework for assessing the value and effectiveness of the adaptation yielded
by adaptive systems. The most common practice of evaluation is experiments.
However, there is not an accurate, clear and agreed framework in which an experiment for the assessing of a student model should be performed. Furthermore, it
is important to not only evaluate but also to ensure that the evaluation uses the
correct methods, since an incorrect method can lead to wrong conclusions (Gena
and Weibelzahl 2007). Besides a well-designed evaluation framework underlines
the success of an approach and its potential value to others (Dempster 2004).

For this reason, the presented evaluation process is performed applying the
evaluation framework PERSIVA (Chrysafiadi and Virvou 2013a), which includes
both questionnaires and observations through experiments. Applying the particular
evaluation framework, either the educational impact (i.e. performance, satisfaction,
change of learners' attitudes) or the adaptation of the personalized and/or adaptive tutoring system is assessed. The evaluation of the educational impact is based
on the Kirkpatrick's model (1979). Furthermore, experiments play a major role in
the particular evaluation method, as they are appropriate for the evaluation of the
effectiveness and successfulness of user models (Chin 2001; Virvou and kabassi
2004).

The remainder of this section is organized as follows. Initially, the evaluation
framework PERSIVA is presented. Then, the evaluation criteria, process and popu-
lation are described. The presentation of the evaluation's results follows. Finally,
the conclusions drawn from the evaluation process are presented.

4.2 The Evaluation Method

4.2.1 The Evaluation Framework PERSIVA

The used evaluation framework PERSIVA (Chrysafiadi and Virvou 2013a) assesses:

- **Students' satisfaction** about the e-learning program. More concretely, information about the feelings, thoughts and satisfaction of the learners' about the adaptivity and effectiveness of the e-learning education environment was gathered.
- **Students' performance** on the knowledge domain.
- The changes that were caused on the **individual state** of the students.
- The **results** of the e-learning program to students' progress.
- The **validity of the conclusions** drawn by the student model concerning the aspects of the students' characteristics.
- The **validity of the adaptation** decision making of the student model.

4.2.2 The Evaluation Criteria

The evaluation of an adaptive educational system is a complex process. It includes evaluation of two different educational aspects. The first aspect concerns the effectiveness of the educational program, and the other aspect concerns the effectiveness of the student model. Therefore, the presented evaluation includes the following two levels:

- **Level 1: Evaluation of the educational impact**. This level includes assessment of the learners' performance and satisfaction; evaluation of the effect of the e-learning program on the behavior and thoughts of students about computer programming and distance learning; evaluation of the effects of the particular e-learning program to students' progress on their further studies and assessment of how it helped the students to learn other programming languages.

- **Level 2: Evaluation of adaptation**. This level is responsible to give answers in the following questions:
  - How satisfied are the learners about the system's adaptive responses to their needs?
  - How important and essential is to model the particular student's characteristics (prior knowledge, knowledge level, type of errors etc.)?
  - How valid are the conclusions drawn by the student model concerning the aspects of the students' characteristics (i.e. their background)?
  - The decisions for adaptivity are valid and effective?

4.2.3 The Evaluation Process

The presented evaluation process includes experiment research and questionnaires. Below, the particular evaluation processes that were used to assess each individual evaluation criteria were described.

- **Learners' general satisfaction**: For gathering this kind of information a questionnaire (Questionnaire A, Appendix B) was used. The questions were close-ended based on Likert scale with five responses ranging from "Very much" (5) to "Not at all" (1). The questions were divided into six categories based on the type of information that were evaluated. The questions of the first category are related to the quality of the content. The questions of the second category concern the quality of instruction. The third and fourth categories concern the friendliness and usefulness of the programming tutoring system correspondingly. The questions of the fifth category are aimed to evaluate the adaptivity of the system. The final question concerns the overall rating of the system.
- **Learners' performance**: The learner's performance is defined by either her/his degree of success in tests and exercises or the times that s/he needed to read a particular domain concept. Therefore, factor p was calculated to assess the learners' performance. In particular, the value of factor p is derived if the learner's degree of success in a particular domain concept is divided with the times that s/he needs to read the particular concept (Eq. 4.1). The better the degree of success is, the higher the value of p is. The lower the value of times that the learner read the concept is, the higher the value of p is. Also, the lower the degree of success is, the lower the value of p is. The higher the value of times that the learner read the concept is, the lower the value of p is. Learners' performance was measured by conducting an experiment with an experimental group (the group of students which used the presented programming tutoring system) and a control group (the group of students which used a similar programming tutoring system from which the presented student model was absent). Factor p was calculated and compared for both groups.

$$p = \frac{\text{degree of success in } C_i}{\text{reading times of } C_i} \qquad (4.1)$$

- **Changes on learners' behavior and thoughts about computer programming**: For gathering this kind of information a questionnaire (Questionnaire B, Appendix B) was used. The questions were close-ended based on Likert scale with five responses ranging from "Very much" (5) to "Not at all" (1).
- **Changes on learners' behavior and thoughts about e-learning**: For gathering this kind of information a questionnaire (Questionnaire C, Appendix B) was used. The questions were close-ended based on Likert scale with five responses ranging from "Very much" (5) to "Not at all" (1).
- **Results on learners' further studies**: The assessment of that criterion is performed with both questionnaire and experiment. In particular, a questionnaire (Questionnaire D, Appendix B), which included five close-ended questions

based on Likert scale with five responses ranging from "Very much" (5) to
"Not at all" (1), was used. The experiment included an experimental group (the
group of students which used the presented programming tutoring system) and
two control groups (students who had not used some programming tutoring system). The average degrees that the students succeeded in two related with programming lessons, which they taught after the use of the programming tutoring
system, were calculated and compared for each of the three groups.

- **Learners' satisfaction about the system's adaptive responses to their
  needs**: For gathering this kind of information a questionnaire (Questionnaire E,
  Appendix B) was used. The questions were close-ended based on Likert scale
  with five responses ranging from "Very much" (5) to "Not at all" (1).

- **The validity of the conclusions drawn by the student model concerning the
  aspects of the students' characteristics**: It is performed assessing the results of
  the system's use in relation with the different backgrounds of the students. More
  concretely, the students, who were used the presented programming tutoring
  system, were divided into three categories according to their background knowledge. These three categories are arts, science fields (other than computer
  science) and computer science related fields. Learners, which have studies in
  the field of human, social, political and educational sciences, belong to arts.
  Learners, who have studies in the field of mathematic, physic, business and
  economic sciences, belong to science fields. Learners, who have studies in
  the field of computer, engineering and system sciences, belong to computer
  science-related fields. It is obvious that students, who belong to the last category, have a previous knowledge on computer programming. Thereby, their
  progress should be better than the students of the other two categories. Also,
  the particular learners should be advised to read a chapter fewer times than the
  others. Furthermore, the learners, who belong to science fields, should have
  a good learning pace, since their studies offer them a way of thinking that is
  close to the programming logic. Whilst, the students of the arts' category should
  be advised to read more times the learning material, until to learn the knowledge domain, since they often have no idea about computer programming.
  Consequently, for each background category, the average reading times of each
  domain concept and the percentage of learners that are advised to return to a
  previous concept in order to revise it were measured. Furthermore, those values
  were compared in order to assess if the conclusions drawn by the system concerning the characteristics of student model are valid.

- **The validity of the adaptation decision making of the student model**: The
  evaluation of the particular criterion includes assessment of the system's adaptation decisions about the return of a learner to a domain concept, which was considered as learner and/or assimilated, in order to revise it and about the inference
  of the system that a learner should not read a particular domain concept at all.
  The particular assessment is performed using a questionnaire and conducting an
  experiment. In particular, a questionnaire (Questionnaire F, Appendix B) with
  close-ended questions based on Likert scale with five responses ranging from
  "Very much" (5) to "Not at all" (1), was used to ask the learners' opinion about

the appropriateness or necessity of the returns to a domain concept, which was considered as learner and/or assimilated, in order to revise it. In addition, learners' performance in a final test was measured and compared for two groups (group A: the group of students which used the presented programming tutoring system, and group B: the group of students which used a similar programming tutoring system with the same organized content, in which the learner chooses if s/he will return to revise the concept that the system indicates her/him and/or if s/he will read or not the suggested concept each time). Furthermore, the percentage of times that a learner needed finally to read a domain concept that the system had advised her/him not to read was measured.

The questionnaires A, E and F (Appendix B) were given to the students after their participation in the training program. However, the learners were asked to complete the questionnaires B, C and D (Appendix B), after, almost, 2 years of their participation in the training program. The reason for that is the fact that the evaluation of the changes on learners' behavior and thoughts and of training program's results on learners' further studies (which correspond to the evaluation levels of behavior and results of the Kirkpatrick's model) need at least a two-year evaluation period (Jeremić et al. 2009). Furthermore, the Questionnaires B and C were, also, given to participants before the use of the systems, in order to compare the answers before and after their participation in the training program.

4.2.4 The Evaluation Population

For the experiment two groups of students were used. Learners of both groups were students of a postgraduate conversion course program in the field of informatics at the University of Piraeus. They had different ages, varying from 22 to 50, and backgrounds. Examples of such backgrounds are physics, mathematics, computer science, education, human and social science. The number of students, which belong to either each age category or background category, is the same for both groups (Table 4.1). The reason for this is the fact that the homogeneity of the experiment's samples simplifies the experiment's performing. Furthermore, 40 % of the learners of the group A had a prior knowledge on computer programming. The learners of group B that knew already a programming language were 45.74 %. The distribution of students' knowledge on other languages for both groups is depicted in Table 4.2.

Table 4.1 Distribution of students' ages and backgrounds

Ages	22–30	31–40	41–50
	68.57 %	22.86 %	8.57 %
Background	Arts	Science (other than computers)	Computer science related
	34.29 %	28.57 %	37.14 %

Table 4.2 Distribution of students' knowledge of other programming languages

Language	Java (%)	Pascal (%)	Basic (%)
Group A	21.43	28.57	50
Group B	22.22	27.78	45

Both groups consisted of 35 students. The students of group A used the presented programming tutoring system for learning the programming language 'C'. The particular, programming tutoring system uses a fuzzy student model, which helps the system to infer about the learner's knowledge level on each domain concept of the learning material and advise the learner not to read a concept at all or return her/him to a previous learnt concept to revise it. The students of group B used a similar educational system from which the fuzzy student model was absent. Both systems had the same knowledge domain, which is divided into 31 concepts, but the system, which was used by the students of group B, delivers the concepts of the learning material in sequence without taking into account how the learner's performance on a domain concept may affect her/his knowledge of another concept. The students, who used that system, had to read all the concepts one time at least. Furthermore, according to the learner's degree of success on tests, the system decided if s/he had to return to revise a concept or if s/he had to be transited to the next section of the learning material. The learners of both groups used the corresponding systems without attending any complementary course on programming, over a period of 6 months.

4.3 Results

4.3.1 Learners' General Satisfaction

Learners' satisfaction about the educational system and program is positive. The average learners' overall rating of satisfaction is 4.38. The level of their satisfaction about the quality of the content, the quality of instruction, the system's usefulness, friendliness and adaptivity is high. The results of the corresponding questionnaire (Questionnaire A, Appendix B) are depicted in Fig. 4.1. This information is easy to collect, but does not tell enough about the learning success.

4.3.2 Learners' Performance

One of the main goals of the system is to adapt dynamically the teaching sequence to the users' individual level of knowledge. In this sense the evaluation's aim is to evaluate the individualization of the teaching rather than evaluating the success of the teaching method alone. In other words, it is evaluated how the student was taught and whether s/he learned successfully rather than just whether s/he learned

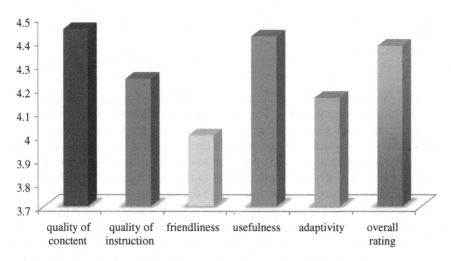

Fig. 4.1 Learners' general satisfaction about the programming tutoring system

successfully. As such, a classical pre-test and post-test methodology could not be sufficient for the aimed evaluation. The value of factor p, which is derived dividing the learner's degree of success in a particular domain concept with the times that s/he needs to read the particular concept, was measured for group A and group B. Then, the two average values were compared to assess the impact of the presented fuzzy student model on the student's performance and progress.

According to Grubišić et al. (2009) experiment used in the e-learning systems' effectiveness evaluation change the independent variable (tutoring strategy) while measuring the depended variable (effects on learning). In the presented work, the experiment has one dependent variable and one independent variable with two values (two independent groups). The dependent variable is the factor p. The independent variable is the groups (group A and group B). Due to the fact that the experiment involves one dependent variable and one independent variable with two levels, the two-independent sample t-test should be used to analyze the experiment's data. The particular statistical test is used to determine whether the different average scores of two groups, represents a real difference between the two populations, or just a chance difference in our samples (Carver and Nash 2009; Norusis 2009). The results of the experiment are depicted in Table 4.3.

However, the different averages scores can be occurred by chance or due to differences on the education, knowledge level and abilities of the learners of the two groups. The Levene's test for equality of variances is used to ensure that the above results were not occurred randomly or due to differences of students' characteristics of the two groups. According to the Levene's test, if the value "Sig." is less than 0.05, then the two variances are significantly different, otherwise the variability in two groups is about the same. The "Sig." value of the experiment is 0.282 (Table 4.4). Therefore, the two variances are approximately equal. Then, the value of "Sig. (2-tailed)" (Table 4.5) is checked in order to infer if the two means

Table 4.3 Group statistics

	Group	N	Mean	Std. deviation	Std. error mean
p	Group A	35	84.8734	16.35326	2.76421
	Group B	35	59.3209	14.17618	2.39621

Table 4.4 Levene's test for equality of variances

		Levene's test for equality of variances		t-test for equality of means	
		F	Sig.	t	df
p	Equal variances assumed	1.174	0.282	6.985	68
	Equal variances not assumed			6.985	66.658

Table 4.5 Independent samples test

		t-test for equality of means				
		Sig. (2-tailed)	Mean difference	Std. error difference	95 % confidence interval of the difference	
					Lower	Upper
p	Equal variances assumed	0.000	25.55257	3.65823	18.25268	32.85246

are statistically different. If this value is greater than 0.05, then there is no statistically difference between the means and the difference is likely due to chance. The experiment's "Sig. (2-tailed)" value is 0. Thereby, the differences between the means are statistically significant and are not a result of chance. Therefore, the presented fuzzy programming tutoring system has a positive effect on the learner's performance and progress.

4.3.3 Changes on Learners' Behavior and Thoughts About Computer Programming

Learners' behavior and thoughts towards the computer programming have positively changed. The questionnaire B that was answered by the learners and the mean of students' answers are displayed in Appendix B. The results of the questionnaire are depicted in Figs. 4.2, 4.3, 4.4 and 4.5. The results show that the state towards computer programming has been improved. The higher improvement is observed for learners with background on arts. It is remarkable the increase of the willingness of the particular learners to be engaged in computer programming projects. Furthermore, the number of learners, who believe that computer programming can facilitate some everyday processes and are motivated to use it in their jobs, has been increased significantly.

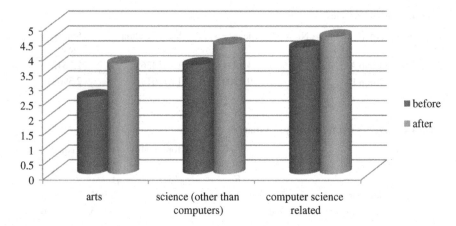

Fig. 4.2 Change's on learners' behavior and thoughts about computer programming

4.3.4 Changes on Learners' Behavior and Thoughts About E-Learning

Learners' behavior and thoughts about distance learning have positively changed. The results of the questionnaire reveal that the students' perception about e-learning has improved. They seem more willing to be involved in e-learning programs. Furthermore, their experience with the web-based programming tutoring system seems to have convinced the learners about the effectiveness and usefulness of the e-learning. The results of the learners' answers to the questionnaire C (Appendix B) are depicted in Figs. 4.6, 4.7, 4.8 and 4.9.

4.3.5 Results on Learners' Further Studies

The results of the e-learning program to the learners' progress on their further studies are satisfactory. The results of the questionnaire reveal that the e-learning program helped the users. The results of the learners' answers to the questionnaire D (Appendix B) are depicted in Fig. 4.10.

However, a supplementary empirical research is essential for gathering this kind of information. For this reason, an experiment was conducted. In the particular experiment, the average degrees that students of three groups succeeded in two computer programming related lessons were calculated. Both lessons were delivered to students after the use of the programming tutoring system. The lesson 1 concerned the computer programming language 'Java'. The lesson 2 concerned the digital signal processing of speech and audio. The latter lesson is not a lesson of computer programming language, but it includes methodologies and techniques of

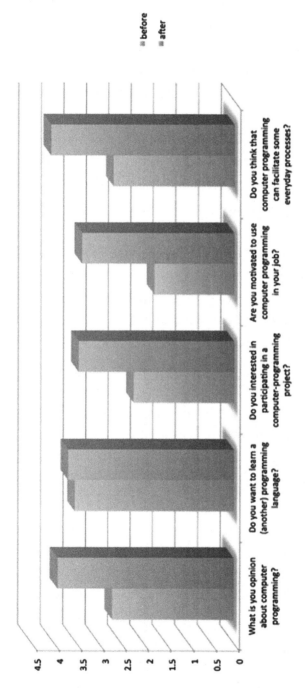

Fig. 4.3 Learners' answers with arts background about computer programing

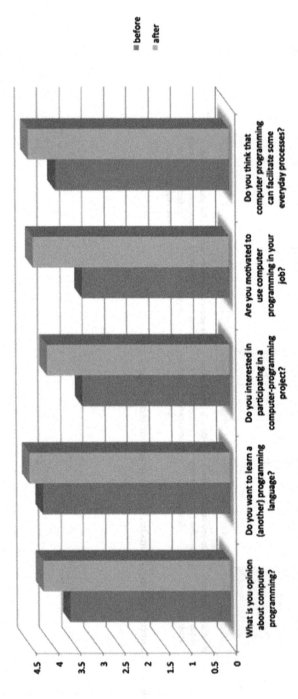

Fig. 4.4 Learners' answers with science (other than computers) background about computer programing

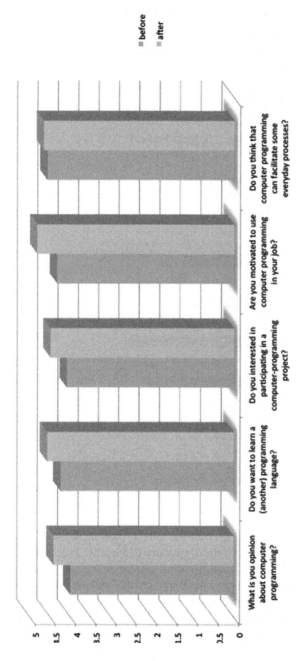

Fig. 4.5 Learners' answers with computer-related science background about computer programing

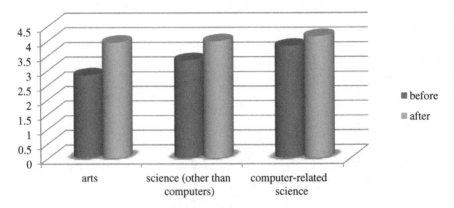

Fig. 4.6 Changes on learners' behavior and thoughts about e-learning

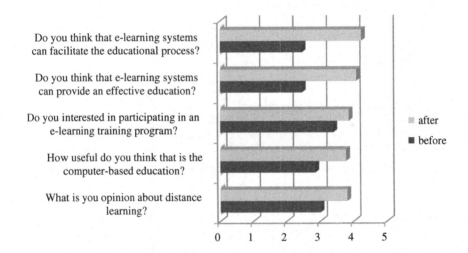

Fig. 4.7 Learners' answers with arts background about e-learning

algorithms and computer programming. Each of the three groups includes 35 students of the department of informatics of the University of Piraeus. The students of group A used the presented programming tutoring system before the teaching of the above lessons, while the students of group B and group C did not used some programming tutoring system. The students' distribution according to their backgrounds is the same for three groups. Particularly, they include 13 students with background on arts, 10 students with background on sciences other than computers, and 12 students with background on computer-related sciences. The results of the experiment were compared in order to assess the effect of the presented web-based programming tutoring system on learners' further studies.

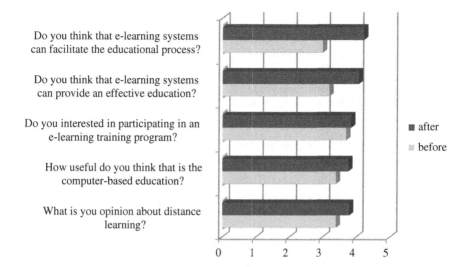

Fig. 4.8 Learners' answers with science (other than computers) background about e-learning

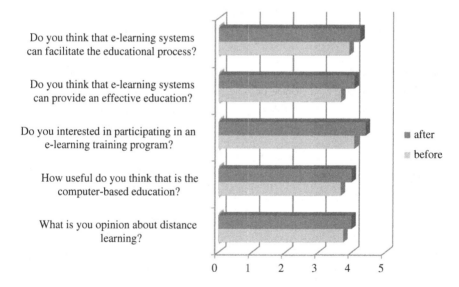

Fig. 4.9 Learners' answers with computer-related science background about e-learning

The particular experiment has two dependent variables and one independent variable with three levels. The dependent variables are the learners' degrees in the computer science lessons. The independent variable is the group of learners and takes three values: group A, group B and group C. Due to the fact that the experiment involves more than one dependent variables at a time, the multivariate analysis of variance (MANOVA) should be used to analyze the experiment's data. The results of the experiment are depicted in Table 4.6. According to the experiment's results,

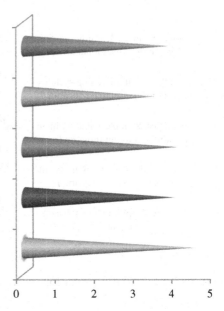

Did the educational software help you in the elaboration of tasks and activities considering your studies?

Did the educational software help you to understand better other lessons of computer science?

Did the educational software help you in your studies?

Did the educational software help you to learn other programming languages?

Did the educational software help you to understand better the logic of programming?

Fig. 4.10   Learners' answers about the results of the e-learning program to their further studies

Table 4.6   Descriptive statistics

	Group	Mean	Std. deviation	N
Grade_on_lesson1	Group A	8.4286	1.00837	35
	Group B	7.3714	1.16533	35
	Group C	7.2857	1.17752	35
	Total	7.6952	1.22572	105
Grade_on_lesson2	Group A	7.5714	0.85011	35
	Group B	6.6571	0.96841	35
	Group C	6.6857	0.96319	35
	Total	6.9714	1.01391	105

the performance of the students, who used the presented web-based tutoring system (group A), is better than the other students. Actually, their performance is better even in lesson of 'digital signal processing of speech and audio', which does not concern computer programming directly.

The homogeneity of covariances and the homogeneity of variances have to be checked to ensure that the different averages scores were not occurred by chance or due to differences on the group's populations. MANOVA checks the homogeneity of covariances conducting the "Box's test of equality of covariance". If the "Sig." value is less than 0.001, then the assumption of homogeneity of covariances is violated. In the particular experiment, that value is 0.344 (Table 4.7). Therefore, there is homogeneity of covariances. Furthermore, MANOVA checks the homogeneity of variances conducting the "Levene's test

Table 4.7 Box's test of equality of covariance matrices

Box's M	6.953
F	1.126
df1	6
df2	259,299.692
Sig.	0.344

of equality of error variances". If the "Sig." value is higher than 0.05, then the variability in experiment's groups is about the same. In the particular experiment, that value is 0.714 for 'grade_on_lesson1' and 0.854 for 'grade_on_lesson2' (Table 4.8). Consequently, both 'grade_on_lesson1' and 'grade_on_lesson2' have homogeneity of variances.

Then, MANOVA conducts the 'Wilks' Lambda test' to determine whether the differences of the means are statistically significant. Lambda varies between 0 and 1. If the "Sig." value of the test is less than 0.0005, then it means that the independent variable has a significant effect on the dependent variable. The lambda value of the independent variable 'group' is 0.760 and the corresponding "Sig." value is 0.000 (Table 4.9). Therefore, learners' performance on lessons 1 and 2 were significantly dependent on the use of the programming tutoring system. Then, tests of "Between-Subjects Effects" determine how the dependent variables differ for the values of the independent variable and if the independent variable has a significant effect on both or only on one dependent variable. From the results that are depicted in Table 4.10, is concluded that the group (so, the use of the programming tutoring system) has a significant effect on both 'grade_on_lesson1' (Sig. = 0.000 < 0.0005) and 'grade_on_lesson2' (Sig. = 0.000 < 0.0005).

Table 4.8 Levene's test of equality of error variances

	F	df1	df2	Sig.
Grade_on_lesson1	0.338	2	102	0.714
Grade_on_lesson2	0.158	2	102	0.854

Table 4.9 Multivariate tests

Effect		Value	F	Hypothesis df	Error df	Sig.
Intercept	Pillai's trace	0.987	3,891.756[a]	2.000	101.000	0.000
	Wilks' lambda	0.013	3,891.756[a]	2.000	101.000	0.000
	Hotelling's trace	77.064	3,891.756[a]	2.000	101.000	0.000
	Roy's largest root	77.064	3,891.756[a]	2.000	101.000	0.000
Group	Pillai's trace	0.241	6.978	4.000	204.000	0.000
	Wilks' lambda	**0.760**	**7.440[a]**	**4.000**	**202.000**	**0.000**
	Hotelling's trace	0.316	7.896	4.000	200.000	0.000
	Roy's largest root	0.314	16.026[b]	2.000	102.000	0.000

[a]Exact Statistic
[b]The statistics is an upper bound on F that yields a lower bound on the significance level

Table 4.10 Tests of between-subjects effects

Source	Dependent variable	Sig.	Partial eta squared	Noncent. parameter	Observed power[*]
Group	Grade_on_lesson1	0.000	0.182	22.621	0.991
	Grade_on_lesson2	0.000	0.177	21.923	0.990

[*]Computed using alpha = .05

Table 4.11 Multiple comparisons

Dependent variable	(I) group	(J) group	Mean difference (I–J)	Std. error	Sig.	95 % confidence interval	
						Lower bound	Upper bound
Grade_on_lesson1	Group A	Group B	1.0571[*]	0.26767	0.000	0.4205	1.6938
		Group C	10.1429[*]	0.26767	0.000	0.5062	1.7795
	Group B	Group A	−1.0571[*]	0.26767	0.000	−1.6938	−0.4205
		Group C	0.0857	0.26767	0.945	−0.5509	0.7223
	Group C	Group A	−1.1429[*]	0.26767	0.000	−1.7795	−0.5062
		Group B	−0.0857	0.26767	0.945	−0.7223	0.5509
Grade_on_lesson2	Group A	Group B	0.9143[*]	0.22204	0.000	0.3862	1.4424
		Group C	0.8857[*]	0.22204	0.000	0.3576	10.4138
	Group B	Group A	−0.9143[*]	0.22204	0.000	−1.4424	−0.3862
		Group C	−0.0286	0.22204	0.991	−0.5567	0.4995
	Group C	Group A	−0.8857[*]	0.22204	0.000	−1.4138	−0.3576
		Group B	0.0286	0.22204	0.991	−0.4995	0.5567

[*]The mean difference is significant at the .05 level

The results of Table 4.11 show that both 'grade_on_lesson1' and 'grade_on_lesson2' are statistical significantly different between either group A and group B (Sig. < 0.0005) or group A and group C (Sig. < 0.0005). However, the dependent variables are not statistical significantly different between group B and group C. Therefore, the use of the presented web-based programming tutoring system has a positive effect on the results of the further studies of the learners.

4.3.6 Learners' Satisfaction About the System's Adaptive Responses to Their Needs

The learners' satisfaction about the system's adaptive responses to their need is very satisfactory. The results of the questionnaire reveal that the programming tutoring system considers the learners needs and reacts dynamically each time to meet them. The results of the learners' answers to the Questionnaire E (Appendix B) are depicted in Fig. 4.11.

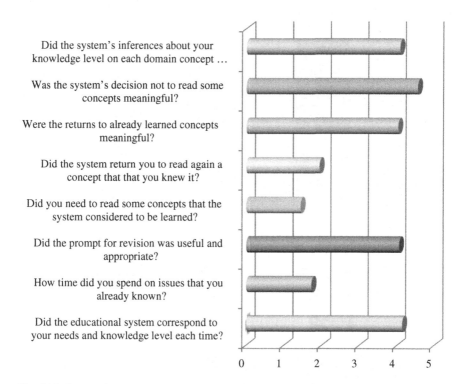

Did the system's inferences about your knowledge level on each domain concept ...

Was the system's decision not to read some concepts meaningful?

Were the returns to already learned concepts meaningful?

Did the system return you to read again a concept that that you knew it?

Did you need to read some concepts that the system considered to be learned?

Did the prompt for revision was useful and appropriate?

How time did you spend on issues that you already known?

Did the educational system correspond to your needs and knowledge level each time?

0   1   2   3   4   5

Fig. 4.11 Learners' answers about the system's adaptive responses to their needs

4.3.7 The Validity of the Conclusions Drawn by the Student Model Concerning the Aspects of the Students' Characteristics

The conclusions that are drawn by the system concerning the aspects of students' characteristics seem to be satisfactory valid. According to the results the system advises the learners with studies on arts fields to read a domain concept more times than the learners who had been involved with the logic of programming before (Fig. 4.12). The lower average times of reading corresponds to the learners with background on computer-related sciences. Furthermore, the system advises the learners with studies on arts to return to a domain concept in order to revise it more times than the other learners (Fig. 4.13). This is evident, since learners with no previous knowledge and experience on computer programming, have difficulty in assimilating the learning material of the particular knowledge domain. The average time of returns to a domain concept for revision is very low for the learners with background on computer-related sciences. It is logical, since the most learners with background on computer-related sciences have already been involved in computer programming and thus it is easy for them to deal with the learning material of the system's knowledge domain.

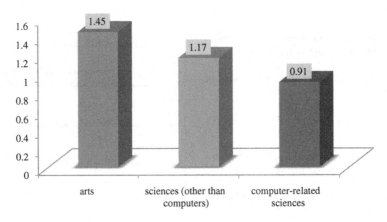

Fig. 4.12 The average times of reading per background category

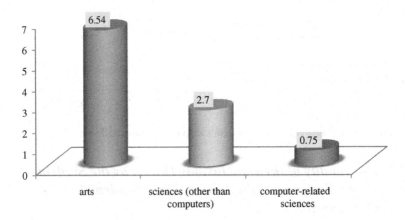

Fig. 4.13 The average times of returns to concepts for revision per background category

In addition, an experiment was conducted to ensure the validity of the conclusions drawn by the student model concerning the aspects of the students' characteristics. In particular, the mean values of times of reading of a domain concept of the learning material were measured for group A and group B. Furthermore, the statistical significance of the difference of the mean values of reading times between the learners of different backgrounds was checked. ANOVA is the statistical method that was chosen for the analysis of the experiment's data. The different mean values of reading times for both groups are presented in Table 4.12.

In the particular experiment, there is homogeneity of variances (the "Sig." value of "Levene's test" is higher than 0.05—Table 4.13). Furthermore, the "Sig." value (Sig. = 0.000 < 0.0005) of the Table 4.14 declares that there is a significant difference on the times of reading between the learners of different backgrounds. Moreover, the comparisons of Table 4.15 show that for group A the mean value of reading times is significantly different between the learners with

Table 4.12 Descriptive statistics of reading times

		Background	Mean	Std. deviation	N
Times_of_ reading	Group A	Arts	1.4533	0.15744	12
		Computer science related	0.9092	0.07805	13
		Science (other than computers)	1.1760	0.12955	10
		Total	1.1720	0.26259	35
	Group B	Arts	1.7325	0.20064	12
		Computer science related	1.2400	0.15138	13
		Science (other than computers)	1.4340	0.19202	10
		Total	1.4643	0.27526	35

Table 4.13 Levene's test of equality of error variances for reading times

F	df1	df2	Sig.
2.223	5	64	0.063

Table 4.14 Tests of between-subjects effects for reading times

Source	Type III sum of squares	df	Mean square	F	Sig.
Corrected model	4.869[a]	5	0.974	40.300	0.000
Intercept	121.273	1	121.273	5,018.842	0.000
Group	1.447	1	1.447	59.894	0.000
Background	3.357	2	1.679	69.473	0.000
Group * background	0.017	2	0.008	0.342	0.712
Error	1.546	64	0.024		
Total	128.040	70			
Corrected total	6.415	69			

Table 4.15 Multiple Comparisons about reading times for group A

(I) background	(J) Background	Mean difference (I–J)	Std. error	Sig.
Arts	Computer science related	0.5441*	0.04988	0.000
	Science (other than computers)	0.2773*	0.05335	0.000
Computer science related	Arts	−0.5441*	0.04988	0.000
	Science (other than computers)	−0.2668*	0.05241	0.000
Science (other than computers)	Arts	−0.2773*	0.05335	0.000
	Computer science related	0.2668*	0.05241	0.000

*The mean difference is significant at the .05 level

background on arts and the learners with background on sciences other than computers (Sig. < 0.0005); between the learners with background on sciences other than computers and the learners with background on computer-related sciences (Sig. < 0.0005); and the learners with background on arts and the learners with

Table 4.16 Multiple comparisons about reading times for group B

(I) Background	(J) Background	Mean difference (I–J)	Std. error	Sig.
Arts	Computer science related	0.4925*	0.07250	0.000
	Science (other than computers)	0.2985*	0.07755	0.002
Computer science related	Arts	−0.4925*	0.07250	0.000
	Science (other than computers)	−0.1940*	0.07618	0.041
Science (other than computers)	Arts	−0.2985*	0.07755	0.002
	Computer science related	0.1940*	0.07618	0.041

*The mean difference is significant at the .05 level

background on computer-related sciences (Sig. < 0.0005). On the other hand, the results of the Table 4.16 show that for group B the mean value of reading times is significantly different only between the learners with background on arts and learners with background on computer-related sciences (Sig. < 0.0005). For group B, the mean value of reading times does not differ significantly between neither the learner with background on arts and learners with background on sciences other than computers (Sig. = 0.002 > 0.0005) nor the learners with background on sciences other than computers and learners with background on computer-related sciences (Sig. = 0.041 > 0.0005). Therefore, the student model of the presented programming tutoring system makes valid conclusions concerning the aspects of the students' characteristics.

4.3.8 The Validity of the Adaptation Decision Making of the Student Model

The results of the validity of the system's decision making were very encouraging. It has been answered that the percentage of time that the learners spent to revise a previous domain concept is at all waste of time. The results of the learners' answers to the questionnaire F (Appendix B) are depicted in Fig. 4.14.

In addition, learners' performance in a final test was measured and compared for two groups (group A: the group of students which used the presented programming tutoring system, and group B: the group of students which used a similar programming tutoring system with the same organized content, in which the learner chooses if s/he will return to revise the concept that the system indicates her/him and/or if s/he will read or not the suggested concept each time). The two-independent sample t-test was used to analyze the experiment's data. The results of the experiment are depicted in Table 4.17. The variability in two groups is about the same (value "Sig." = 0.106 > 0.05 Table 4.18). Therefore, the different average scores did not occur by chance or due to differences on the education, knowledge level and abilities of the learners of the two groups. Furthermore, the differences between the mean values are statistically significant; since the value of

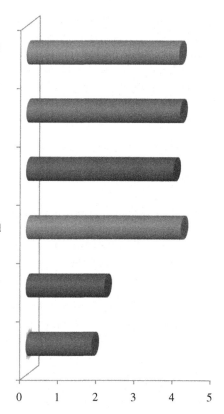

Did the returns to a previous domain concept
for revision help you to learn computer
programming better?

Were the returns to already learned concepts
meaningful?

Did the returns to a previous domain concept
correspond to your need for revision?

Did the prompt for revision was useful and
appropriate?

How many times the returns to a previous
read concept for revision concerned
concepts that you actually knew?

Were the returns to a previous learned
concept for revision a waste of time?

Fig. 4.14 Learners' answers about about the appropriateness or necessity of the returns to a domain concept for revision

Table 4.17 Mean values of performance

	Group	N	Mean	Std. deviation	Std. error mean
Performance	Group A	35	90.1463	3.78963	0.64056
	Group B	35	82.7143	5.30023	0.89590

Table 4.18 Independent samples test for performance

		Levene's test for equality of variances		t-test for equality of means				
		F	Sig.	t	df	Sig. (2-tailed)	Mean difference	Std. error difference
Performance	Equal variances assumed	2.676	0.106	6.748	68	0.000	7.43200	1.10135
	Equal variances not assumed			6.748	61.56	0.000	7.43200	1.10135

"Sig. (2-tailed)" (Table 4.18) is lower than 0.05. Consequently, the system's decisions about adaptive responses to the learner's needs are valid, since learners seem to assimilate the learning material and succeed a better performance finally.

In addition, the percentage of times that a learner needed, finally, to read a domain concept that the system had advised her/him not to read it is 7.14 %. This percentage is sufficiently satisfactory to be able to lead to the conclusion that the decisions, which are made by the system based on the student model, are valid.

4.4 Conclusions

The system's evaluation revealed that the combination of fuzzy sets with overlay model and stereotypes contributes, significantly, to the adaptation of the learning process to the learning pace of each individual learner. The results of the evaluation demonstrated learning improvements and successful adaptation to students' needs. In particular, the learners' overall rating of the presented web-based programming tutoring system is very high. The participant learners were very satisfied with the quality of content and quality, with the tutoring system's friendliness and usefulness, and with the system's adaptive responses to their needs. Furthermore, the integration of the presented novel fuzzy student model into the programming tutoring system improved significantly the student's performance. Also, the learners, who used the presented e-learning tutoring system, obtained a more positive state and behavior towards computer programming and distance learning. The assessment results showed, also, that the e-learning program helped learners to their further studies satisfactory. In addition, the evaluation results revealed that the presented novel fuzzy approach of student modeling improves the efficiency of the adaptation of the instructional process. In particular, both, the conclusions that are drawn by the system concerning the aspects of students' characteristics and the adaptation decision-making were valid. The system advises the student model and adapts instantly the sequence of learning lessons concerning the students' characteristics with success. Consequently, the presented novel hybrid student model contributes significantly to the adaptation process and helps the system to provide a personalized and effective educational process for learning of computer programming.

Conclusions and Discussion

Applicability of the Novel Fuzzy Student Modeling Approach

The target of this book was to present a survey of student modeling approaches and fuzzy logic application and a novel approach that combines fuzzy logic techniques for offering individualized instruction and personalized support in adaptive educational systems. The presented student modeling approach performs individualized instruction, adapting the delivery of the knowledge domain to the individual learner's learning needs and pace. The presented approach models either how learning progresses or how the student's knowledge can decrease. It automatically models the learning or forgetting process of a student. In particular, the system keeps track of cognitive state transitions of learners with respect to their progress or no-progress. Thereby, it reveals if a student learns or not, if s/he forgets and reasons these states. Therefore, the system allows each individual learner to complete the e-learning course at her/his own pace, taking decisions about the concepts of the learning material that have to be delivered to them, the concepts that need revision and the concepts that are known and do not need further reading. In this way, the system helps learners to save time and effort during the learning process.

The presented novel approach was fully implemented and evaluated. Particularly, an original integrated environment for personalized e-training in programming and the language 'C' was developed. This system was used by the students of a postgraduate program in the field of Informatics in the University of Piraeus, Greece. The evaluation was based on close-ended questionnaires and experimental research. The results of the evaluation were very encouraging. They demonstrated that the system models the student's cognitive state and adapts dynamically to her/his individual needs by scheduling the sequence of lessons on the fly, allowing her/him to complete the e-training course at her/his own pace and according to her/his ability. The encouraging results motivate the implementation of the presented novel approach of tutoring system that combines fuzzy techniques to educational environments of knowledge domains other than computer programming.

© Springer International Publishing Switzerland 2015
K. Chrysafiadi and M. Virvou, *Advances in Personalized Web-Based Education*,
Intelligent Systems Reference Library 78, DOI 10.1007/978-3-319-12895-5

Contribution to Knowledge Domain Representation

The operation of the system is based on the knowledge domain representation that is implemented through a Fuzzy Related-Concepts Network (FR-CN). This kind of knowledge domain representation helps to manage to represent either the order in which the domain concepts of the learning material have to be taught and organized, or the knowledge dependencies that exist among the domain concepts. This is significant because the knowledge level of a domain concept increases or decreases due to changes on the knowledge level of a related domain concept. The design of the learning material and the definition of the individual domain concepts that it includes, are based on the knowledge and experience of domain experts. Furthermore, the contribution of domain experts is significant for the definition of the knowledge dependencies that exist among the domain concepts of the learning material and their "strength o impact" on each other.

The particular knowledge domain representation approach helps the system to recognize either the domain concepts that are already partly or completely known for a learner or the domain concepts that s/he has forgot, taking into account the learner's knowledge level of the related concepts of the learning material. As a consequence, the presented knowledge domain representation approach contributes to the improvement of the navigation support that an adaptive and/or personalized learning system provides. Furthermore, the presented approach represents the knowledge domain in a more realistic way. It constitutes a prototype for an adaptive and/or personalized tutoring system for delivering the learning material to each individual learner dynamically, taking into account her/his learning needs and different learning pace.

Contribution to Student Modeling

The target of this book was to show how fuzzy sets can be combined with other student modeling techniques to promote adaptivity and personalization in educational applications. The evaluation of the novel approach, which was presented in this book, revealed that the incorporation of fuzzy techniques into the student model contributes significantly to the adaptation of the learning process to the learning pace of each individual learner. The presented novel fuzzy student modeling approach allows the system to identify the appropriate domain concepts that correspond to each individual learner's knowledge level and educational needs. Therefore, it improves the efficiency of the adaptivity of the instructional process.

The presented fuzzy student modeling approach models automatically the learning or forgetting process of a student. In particular, it helps the learners that already know some concepts of the teaching material to save time and effort during the learning process. Furthermore, the presented approach helps the system to recognize which domain concepts of the learning material that the student has already learned in previous interactions, s/he has forgotten and adapts the presentation of material accordingly.

The ability of the presented novel fuzzy student model to recognize the alterations of the student's learning states and dynamically adapt the presentation of the learning material accordingly, renders the particular student modeling approach a novel generic tool for adaptive learning. The gain from the presented approach is significant as fuzzy logic can be used in combination with overlay and stereotype models to provide adaptivity and personalization in other interactive systems in addition to educational applications. The application of this innovative approach is possible where the user's changeable state and/or preferences are affected by the existing dependencies among the system's elements (like concepts, preferences, events, choices).

Contribution to Programming Tutoring Systems

Programming tutoring systems teaches computer programming to learners providing adaptivity. Mainly, these systems adapt the learning process dynamically to the student's knowledge level and needs. However, they do not consider how the learner's performance in a domain concept affects the learner's knowledge level of other related domain concepts of the learning material. Consequently, the gain of the presented approach is the modeling of the learning or forgetting process of a student that gives the ability to the system to adapt dynamically to each individual learner's needs by scheduling the sequence of lessons instantly.

The presented approach allows the tutoring system to recognize either the learner's knowledge level, or the alterations that occur in the learner's knowledge of a domain concept. Then, the system updates the student's knowledge level of the domain concepts of the learning material that are related with the concept that the student has learnt or forgotten. Therefore, the presented novel approach allows the programming tutoring system to model either the possible increase or decrease of the learner's knowledge. Furthermore, the presented approach introduces the reasoning of errors that are related with other programming languages. In particular, each time the system checks if the learner's errors were due to possible confusion with features of another previously-known programming language. In this way, the system allows each learner to complete the e-learning course at their own pace, taking decisions about which concepts have to be delivered, which concepts need revision and which concepts are known and do not need further reading.

Contribution to Fuzzy Logic

The gain from the novel fuzzy student model is that the learner's knowledge level is represented in a more realistic way. It considers the fact that the learner's knowledge level is a moving target and models automatically the learning and forgetting process. In particular, the presented fuzzy student modeling approach helps the tutoring system to recognize when a new domain concept is completely unknown to the learner, when it is fully learned or assimilated, or when it is partly known due to the learner having previous related knowledge. This is achieved using fuzzy sets for describing how well each individual domain concept is known or learned and a set of fuzzy rules, which are responsible for the update of the overall learner's knowledge level after any change of her/his knowledge of a particular domain of the learning material.

The application of this approach is not limited to adaptive instruction, but it can also be used in other systems with changeable user states, such as e-shops, where consumers' preferences change over the time and affect one another. For example, it can be used to reduce customer information overload by recommending products that are likely to be of interest to them, considering their preferences and the dependencies that exist between products' choices (in accordance to e-learning, users' preferences correspond to users' knowledge level and products' choices correspond to the domain concepts). Therefore, the particular novel fuzzy approach constitutes a novel generic fuzzy tool, which offers dynamic adaptation to users' needs and preferences of adaptive systems.

Appendix A: The Matrixes of the System's FR-CN

Table A.1 The ORDER matrix of the FR-CN of Fig. 2.11

(a)

		1	2	3	4	5	6	7	8	9	10	11	12	13	14	15
		C_1	$C_{1.1}$	$C_{1.2}$	$C_{1.3}$	$C_{1.4}$	$C_{1.5}$	$C_{1.6}$	$C_{1.7}$	C_2	$C_{2.1}$	C_3	$C_{3.1}$	$C_{3.2}$	$C_{3.2.1}$	$C_{3.3}$
1	C_1	0	0	0	0	0	0	0	0	1	1	1	1	1	1	1
2	$C_{1.1}$	0	0	0	0	0	0	0	0	1	1	1	1	1	1	1
3	$C_{1.2}$	0	0	0	0	0	0	0	0	1	1	1	1	1	1	1
4	$C_{1.3}$	0	0	0	0	0	0	0	0	1	1	1	1	1	1	1
5	$C_{1.4}$	0	0	0	0	0	0	0	0	1	1	1	1	1	1	1
6	$C_{1.5}$	0	0	0	0	0	0	0	0	1	1	1	1	1	1	1
7	$C_{1.6}$	0	0	0	0	0	0	0	0	1	1	1	1	1	1	1
8	$C_{1.7}$	0	0	0	0	0	0	0	0	1	1	1	1	1	1	1
9	C_2	0	0	0	0	0	0	0	0	0	0	1	1	1	1	1
10	$C_{2.1}$	0	0	0	0	0	0	0	0	0	0	1	1	1	1	1
11	C_3	0	0	0	0	0	0	0	0	0	0	0	0	0	0	0
12	$C_{3.1}$	0	0	0	0	0	0	0	0	0	0	0	0	1	0	0
13	$C_{3.2}$	0	0	0	0	0	0	0	0	0	0	0	0	0	0	0
14	$C_{3.2.1}$	0	0	0	0	0	0	0	0	0	0	0	0	0	0	0
15	$C_{3.3}$	0	0	0	0	0	0	0	0	0	0	0	0	0	0	0
16	C_4	0	0	0	0	0	0	0	0	0	0	0	0	0	0	0
17	$C_{4.1}$	0	0	0	0	0	0	0	0	0	0	0	0	0	0	0
18	$C_{4.2}$	0	0	0	0	0	0	0	0	0	0	0	0	0	0	0
19	$C_{4.3}$	0	0	0	0	0	0	0	0	0	0	0	0	0	0	0
20	$C_{4.4}$	0	0	0	0	0	0	0	0	0	0	0	0	0	0	0

(continued)

© Springer International Publishing Switzerland 2015
K. Chrysafiadi and M. Virvou, *Advances in Personalized Web-Based Education*,
Intelligent Systems Reference Library 78, DOI 10.1007/978-3-319-12895-5

Table A.1 (continued)

(a)

		1	2	3	4	5	6	7	8	9	10	11	12	13	14	15
21	$C_{4.5}$	0	0	0	0	0	0	0	0	0	0	0	0	0	0	0
22	C_5	0	0	0	0	0	0	0	0	0	0	0	0	0	0	0
23	$C_{5.1}$	0	0	0	0	0	0	0	0	0	0	0	0	0	0	0
24	$C_{5.2}$	0	0	0	0	0	0	0	0	0	0	0	0	0	0	0
25	$C_{5.3}$	0	0	0	0	0	0	0	0	0	0	0	0	0	0	0
26	$C_{5.4}$	0	0	0	0	0	0	0	0	0	0	0	0	0	0	0
27	$C_{5.5}$	0	0	0	0	0	0	0	0	0	0	0	0	0	0	0
28	$C_{5.6}$	0	0	0	0	0	0	0	0	0	0	0	0	0	0	0
29	C_6	0	0	0	0	0	0	0	0	0	0	0	0	0	0	0
30	$C_{6.1}$	0	0	0	0	0	0	0	0	0	0	0	0	0	0	0
31	$C_{6.2}$	0	0	0	0	0	0	0	0	0	0	0	0	0	0	0
32	$C_{6.3}$	0	0	0	0	0	0	0	0	0	0	0	0	0	0	0
33	$C_{6.4}$	0	0	0	0	0	0	0	0	0	0	0	0	0	0	0
34	$C_{6.5}$	0	0	0	0	0	0	0	0	0	0	0	0	0	0	0
35	$C_{6.6}$	0	0	0	0	0	0	0	0	0	0	0	0	0	0	0
36	$C_{6.7}$	0	0	0	0	0	0	0	0	0	0	0	0	0	0	0
37	C_7	0	0	0	0	0	0	0	0	0	0	0	0	0	0	0
38	$C_{7.1}$	0	0	0	0	0	0	0	0	0	0	0	0	0	0	0

(b)

		16	17	18	19	20	21	22	23	24	25	26	27	28
		C_4	$C_{4.1}$	$C_{4.2}$	$C_{4.3}$	$C_{4.4}$	$C_{4.5}$	C_5	$C_{5.1}$	$C_{5.2}$	$C_{5.3}$	$C_{5.4}$	$C_{5.5}$	$C_{5.6}$
1	C_1	1	1	1	1	1	1	1	1	1	1	1	1	1
2	$C_{1.1}$	1	1	1	1	1	1	1	1	1	1	1	1	1
3	$C_{1.2}$	1	1	1	1	1	1	1	1	1	1	1	1	1
4	$C_{1.3}$	1	1	1	1	1	1	1	1	1	1	1	1	1
5	$C_{1.4}$	1	1	1	1	1	1	1	1	1	1	1	1	1
6	$C_{1.5}$	1	1	1	1	1	1	1	1	1	1	1	1	1
7	$C_{1.6}$	1	1	1	1	1	1	1	1	1	1	1	1	1
8	$C_{1.7}$	1	1	1	1	1	1	1	1	1	1	1	1	1
9	C_2	1	1	1	1	1	1	1	1	1	1	1	1	1
10	$C_{2.1}$	1	1	1	1	1	1	1	1	1	1	1	1	1
11	C_3	1	1	1	1	1	1	1	1	1	1	1	1	1
12	$C_{3.1}$	1	1	1	1	1	1	1	1	1	1	1	1	1
13	$C_{3.2}$	1	1	1	1	1	1	1	1	1	1	1	1	1
14	$C_{3.2.1}$	1	1	1	1	1	1	1	1	1	1	1	1	1
15	$C_{3.3}$	1	1	1	1	1	1	1	1	1	1	1	1	1
16	C_4	0	0	0	0	0	0	1	1	1	1	1	1	1
17	$C_{4.1}$	0	0	1	0	0	0	1	1	1	1	1	1	1
18	$C_{4.2}$	0	0	0	0	1	0	1	1	1	1	1	1	1
19	$C_{4.3}$	0	0	0	0	0	0	1	1	1	1	1	1	1
20	$C_{4.4}$	0	0	0	0	0	0	1	1	1	1	1	1	1

(continued)

Table A.1 (continued)

(b)

		16	17	18	19	20	21	22	23	24	25	26	27	28
21	$C_{4.5}$	0	0	0	0	0	0	1	1	1	1	1	1	1
22	C_5	0	0	0	0	0	0	0	0	0	0	0	0	0
23	$C_{5.1}$	0	0	0	0	0	0	0	0	0	0	0	0	0
24	$C_{5.2}$	0	0	0	0	0	0	0	0	0	0	0	0	0
25	$C_{5.3}$	0	0	0	0	0	0	0	0	0	0	0	0	0
26	$C_{5.4}$	0	0	0	0	0	0	0	0	0	0	0	0	0
27	$C_{5.5}$	0	0	0	0	0	0	0	0	0	0	0	0	0
28	$C_{5.6}$	0	0	0	0	0	0	0	0	0	0	0	0	0
29	C_6	0	0	0	0	0	0	0	0	0	0	0	0	0
30	$C_{6.1}$	0	0	0	0	0	0	0	0	1	0	0	0	0
31	$C_{6.2}$	0	0	0	0	0	0	0	0	0	0	1	0	0
32	$C_{6.3}$	0	0	0	0	0	0	0	0	0	0	1	0	0
33	$C_{6.4}$	0	0	0	0	0	0	0	0	0	0	0	0	0
34	$C_{6.5}$	0	0	0	0	0	0	0	0	0	0	0	0	0
35	$C_{6.6}$	0	0	0	0	0	0	0	0	0	0	0	0	0
36	$C_{6.7}$	0	0	0	0	0	0	0	0	0	0	0	0	0
37	C_7	0	0	0	0	0	0	0	0	0	0	0	0	0
38	$C_{7.1}$	0	0	0	0	0	0	0	0	0	0	0	0	0

(c)

		29	30	31	32	33	34	35	36	37	38
		C_6	$C_{6.1}$	$C_{6.2}$	$C_{6.3}$	$C_{6.4}$	$C_{6.5}$	$C_{6.6}$	$C_{6.7}$	C_7	$C_{7.1}$
1	C_1	1	1	1	1	1	1	1	1	1	1
2	$C_{1.1}$	1	1	1	1	1	1	1	1	1	1
3	$C_{1.2}$	1	1	1	1	1	1	1	1	1	1
4	$C_{1.3}$	1	1	1	1	1	1	1	1	1	1
5	$C_{1.4}$	1	1	1	1	1	1	1	1	1	1
6	$C_{1.5}$	1	1	1	1	1	1	1	1	1	1
7	$C_{1.6}$	1	1	1	1	1	1	1	1	1	1
8	$C_{1.7}$	1	1	1	1	1	1	1	1	1	1
9	C_2	1	1	1	1	1	1	1	1	1	1
10	$C_{2.1}$	1	1	1	1	1	1	1	1	1	1
11	C_3	1	1	1	1	1	1	1	1	1	1
12	$C_{3.1}$	1	1	1	1	1	1	1	1	1	1
13	$C_{3.2}$	1	1	1	1	1	1	1	1	1	1
14	$C_{3.2.1}$	1	1	1	1	1	1	1	1	1	1
15	$C_{3.3}$	1	1	1	1	1	1	1	1	1	1
16	C_4	1	1	1	1	1	1	1	1	1	1
17	$C_{4.1}$	1	1	1	1	1	1	1	1	1	1
18	$C_{4.2}$	1	1	1	1	1	1	1	1	1	1
19	$C_{4.3}$	1	1	1	1	1	1	1	1	1	1

(continued)

Table A.1 (continued)

(c)

		29	30	31	32	33	34	35	36	37	38
20	$C_{4.4}$	1	1	1	1	1	1	1	1	1	1
21	$C_{4.5}$	1	1	1	1	1	1	1	1	1	1
22	C_5	1	1	1	1	1	1	1	1	1	1
23	$C_{5.1}$	1	1	1	1	1	1	1	1	1	1
24	$C_{5.2}$	1	1	1	1	1	1	1	1	1	1
25	$C_{5.3}$	1	1	1	1	1	1	1	1	1	1
26	$C_{5.4}$	1	1	1	1	1	1	1	1	1	1
27	$C_{5.5}$	1	1	1	1	1	1	1	1	1	1
28	$C_{5.6}$	1	1	1	1	1	1	1	1	1	1
29	C_6	0	0	0	0	0	0	0	0	1	1
30	$C_{6.1}$	0	0	0	0	0	1	0	0	1	1
31	$C_{6.2}$	0	0	0	0	0	0	0	0	1	1
32	$C_{6.3}$	0	0	0	0	0	0	0	0	1	1
33	$C_{6.4}$	0	0	0	0	0	0	0	0	1	1
34	$C_{6.5}$	0	0	0	0	0	0	1	1	1	1
35	$C_{6.6}$	0	0	0	0	0	0	0	0	1	1
36	$C_{6.7}$	0	0	0	0	0	0	0	0	1	1
37	C_7	0	0	0	0	0	0	0	0	0	0
38	$C_{7.1}$	0	0	0	0	0	0	0	0	0	0

Table A.2 The matrix PART of the FR-CN of Fig. 2.11

(a)

		1	2	3	4	5	6	7	8	9	10	11	12	13	14	15
		C_1	$C_{1.1}$	$C_{1.2}$	$C_{1.3}$	$C_{1.4}$	$C_{1.5}$	$C_{1.6}$	$C_{1.7}$	C_2	$C_{2.1}$	C_3	$C_{3.1}$	$C_{3.2}$	$C_{3.2.1}$	$C_{3.3}$
1	C_1	0	0	0	0	0	0	0	0	0	0	0	0	0	0	0
2	$C_{1.1}$	1	0	0	0	0	0	0	0	0	0	0	0	0	0	0
3	$C_{1.2}$	1	0	0	0	0	0	0	0	0	0	0	0	0	0	0
4	$C_{1.3}$	1	0	0	0	0	0	0	0	0	0	0	0	0	0	0
5	$C_{1.4}$	1	0	0	0	0	0	0	0	0	0	0	0	0	0	0
6	$C_{1.5}$	1	0	0	0	0	0	0	0	0	0	0	0	0	0	0
7	$C_{1.6}$	1	0	0	0	0	0	0	0	0	0	0	0	0	0	0
8	$C_{1.7}$	1	0	0	0	0	0	0	0	0	0	0	0	0	0	0
9	C_2	0	0	0	0	0	0	0	0	0	0	0	0	0	0	0
10	$C_{2.1}$	0	0	0	0	0	0	0	0	1	0	0	0	0	0	0
11	C_3	0	0	0	0	0	0	0	0	0	0	0	0	0	0	0
12	$C_{3.1}$	0	0	0	0	0	0	0	0	0	0	1	0	0	0	0

(continued)

Table A.2 (continued)

(a)

		1	2	3	4	5	6	7	8	9	10	11	12	13	14	15
13	$C_{3.2}$	0	0	0	0	0	0	0	0	0	0	1	0	0	0	0
14	$C_{3.2.1}$	0	0	0	0	0	0	0	0	0	0	1	0	1	0	0
15	$C_{3.3}$	0	0	0	0	0	0	0	0	0	0	1	0	0	0	0
16	C_4	0	0	0	0	0	0	0	0	0	0	0	0	0	0	0
17	$C_{4.1}$	0	0	0	0	0	0	0	0	0	0	0	0	0	0	0
18	$C_{4.2}$	0	0	0	0	0	0	0	0	0	0	0	0	0	0	0
19	$C_{4.3}$	0	0	0	0	0	0	0	0	0	0	0	0	0	0	0
20	$C_{4.4}$	0	0	0	0	0	0	0	0	0	0	0	0	0	0	0
21	$C_{4.5}$	0	0	0	0	0	0	0	0	0	0	0	0	0	0	0
22	C_5	0	0	0	0	0	0	0	0	0	0	0	0	0	0	0
23	$C_{5.1}$	0	0	0	0	0	0	0	0	0	0	0	0	0	0	0
24	$C_{5.2}$	0	0	0	0	0	0	0	0	0	0	0	0	0	0	0
25	$C_{5.3}$	0	0	0	0	0	0	0	0	0	0	0	0	0	0	0
26	$C_{5.4}$	0	0	0	0	0	0	0	0	0	0	0	0	0	0	0
27	$C_{5.5}$	0	0	0	0	0	0	0	0	0	0	0	0	0	0	0
28	$C_{5.6}$	0	0	0	0	0	0	0	0	0	0	0	0	0	0	0
29	C_6	0	0	0	0	0	0	0	0	0	0	0	0	0	0	0
30	$C_{6.1}$	0	0	0	0	0	0	0	0	0	0	0	0	0	0	0
31	$C_{6.2}$	0	0	0	0	0	0	0	0	0	0	0	0	0	0	0
32	$C_{6.3}$	0	0	0	0	0	0	0	0	0	0	0	0	0	0	0
33	$C_{6.4}$	0	0	0	0	0	0	0	0	0	0	0	0	0	0	0
34	$C_{6.5}$	0	0	0	0	0	0	0	0	0	0	0	0	0	0	0
35	$C_{6.6}$	0	0	0	0	0	0	0	0	0	0	0	0	0	0	0
36	$C_{6.7}$	0	0	0	0	0	0	0	0	0	0	0	0	0	0	0
37	C_7	0	0	0	0	0	0	0	0	0	0	0	0	0	0	0
38	$C_{7.1}$	0	0	0	0	0	0	0	0	0	0	0	0	0	0	0

(b)

		16	17	18	19	20	21	22	23	24	25	26	27	28
		C_4	$C_{4.1}$	$C_{4.2}$	$C_{4.3}$	$C_{4.4}$	$C_{4.5}$	C_5	$C_{5.1}$	$C_{5.2}$	$C_{5.3}$	$C_{5.4}$	$C_{5.5}$	$C_{5.6}$
1	C_1	0	0	0	0	0	0	0	0	0	0	0	0	0
2	$C_{1.1}$	0	0	0	0	0	0	0	0	0	0	0	0	0
3	$C_{1.2}$	0	0	0	0	0	0	0	0	0	0	0	0	0
4	$C_{1.3}$	0	0	0	0	0	0	0	0	0	0	0	0	0
5	$C_{1.4}$	0	0	0	0	0	0	0	0	0	0	0	0	0
6	$C_{1.5}$	0	0	0	0	0	0	0	0	0	0	0	0	0
7	$C_{1.6}$	0	0	0	0	0	0	0	0	0	0	0	0	0
8	$C_{1.7}$	0	0	0	0	0	0	0	0	0	0	0	0	0
9	C_2	0	0	0	0	0	0	0	0	0	0	0	0	0
10	$C_{2.1}$	0	0	0	0	0	0	0	0	0	0	0	0	0
11	C_3	0	0	0	0	0	0	0	0	0	0	0	0	0
12	$C_{3.1}$	0	0	0	0	0	0	0	0	0	0	0	0	0

(continued)

Table A.2 (continued)

(b)

		16	17	18	19	20	21	22	23	24	25	26	27	28
13	$C_{3.2}$	0	0	0	0	0	0	0	0	0	0	0	0	0
14	$C_{3.2.1}$	0	0	0	0	0	0	0	0	0	0	0	0	0
15	$C_{3.3}$	0	0	0	0	0	0	0	0	0	0	0	0	0
16	C_4	0	0	0	0	0	0	0	0	0	0	0	0	0
17	$C_{4.1}$	1	0	0	0	0	0	0	0	0	0	0	0	0
18	$C_{4.2}$	1	0	0	0	0	0	0	0	0	0	0	0	0
19	$C_{4.3}$	1	0	0	0	0	0	0	0	0	0	0	0	0
20	$C_{4.4}$	1	0	0	0	0	0	0	0	0	0	0	0	0
21	$C_{4.5}$	1	0	0	0	0	0	0	0	0	0	0	0	0
22	C_5	0	0	0	0	0	0	0	0	0	0	0	0	0
23	$C_{5.1}$	0	0	0	0	0	0	1	0	0	0	0	0	0
24	$C_{5.2}$	0	0	0	0	0	0	1	0	0	0	0	0	0
25	$C_{5.3}$	0	0	0	0	0	0	1	0	0	0	0	0	0
26	$C_{5.4}$	0	0	0	0	0	0	1	0	0	0	0	0	0
27	$C_{5.5}$	0	0	0	0	0	0	1	0	0	0	0	0	0
28	$C_{5.6}$	0	0	0	0	0	0	1	0	0	0	0	0	0
29	C_6	0	0	0	0	0	0	0	0	0	0	0	0	0
30	$C_{6.1}$	0	0	0	0	0	0	0	0	0	0	0	0	0
31	$C_{6.2}$	0	0	0	0	0	0	0	0	0	0	0	0	0
32	$C_{6.3}$	0	0	0	0	0	0	0	0	0	0	0	0	0
33	$C_{6.4}$	0	0	0	0	0	0	0	0	0	0	0	0	0
34	$C_{6.5}$	0	0	0	0	0	0	0	0	0	0	0	0	0
35	$C_{6.6}$	0	0	0	0	0	0	0	0	0	0	0	0	0
36	$C_{6.7}$	0	0	0	0	0	0	0	0	0	0	0	0	0
37	C_7	0	0	0	0	0	0	0	0	0	0	0	0	0
38	$C_{7.1}$	0	0	0	0	0	0	0	0	0	0	0	0	0

(c)

		29	30	31	32	33	34	35	36	37	38
		C_6	$C_{6.1}$	$C_{6.2}$	$C_{6.3}$	$C_{6.4}$	$C_{6.5}$	$C_{6.6}$	$C_{6.7}$	C_7	$C_{7.1}$
1	C_1	0	0	0	0	0	0	0	0	0	0
2	$C_{1.1}$	0	0	0	0	0	0	0	0	0	0
3	$C_{1.2}$	0	0	0	0	0	0	0	0	0	0
4	$C_{1.3}$	0	0	0	0	0	0	0	0	0	0
5	$C_{1.4}$	0	0	0	0	0	0	0	0	0	0
6	$C_{1.5}$	0	0	0	0	0	0	0	0	0	0
7	$C_{1.6}$	0	0	0	0	0	0	0	0	0	0
8	$C_{1.7}$	0	0	0	0	0	0	0	0	0	0
9	C_2	0	0	0	0	0	0	0	0	0	0
10	$C_{2.1}$	0	0	0	0	0	0	0	0	0	0
11	C_3	0	0	0	0	0	0	0	0	0	0

(continued)

Table A.2 (continued)

(c)

		29	30	31	32	33	34	35	36	37	38
12	$C_{3.1}$	0	0	0	0	0	0	0	0	0	0
13	$C_{3.2}$	0	0	0	0	0	0	0	0	0	0
14	$C_{3.2.1}$	0	0	0	0	0	0	0	0	0	0
15	$C_{3.3}$	0	0	0	0	0	0	0	0	0	0
16	C_4	0	0	0	0	0	0	0	0	0	0
17	$C_{4.1}$	0	0	0	0	0	0	0	0	0	0
18	$C_{4.2}$	0	0	0	0	0	0	0	0	0	0
19	$C_{4.3}$	0	0	0	0	0	0	0	0	0	0
20	$C_{4.4}$	0	0	0	0	0	0	0	0	0	0
21	$C_{4.5}$	0	0	0	0	0	0	0	0	0	0
22	C_5	0	0	0	0	0	0	0	0	0	0
23	$C_{5.1}$	0	0	0	0	0	0	0	0	0	0
24	$C_{5.2}$	0	0	0	0	0	0	0	0	0	0
25	$C_{5.3}$	0	0	0	0	0	0	0	0	0	0
26	$C_{5.4}$	0	0	0	0	0	0	0	0	0	0
27	$C_{5.5}$	0	0	0	0	0	0	0	0	0	0
28	$C_{5.6}$	0	0	0	0	0	0	0	0	0	0
29	C_6	0	0	0	0	0	0	0	0	0	0
30	$C_{6.1}$	1	0	0	0	0	0	0	0	0	0
31	$C_{6.2}$	1	0	0	0	0	0	0	0	0	0
32	$C_{6.3}$	1	0	0	0	0	0	0	0	0	0
33	$C_{6.4}$	1	0	0	0	0	0	0	0	0	0
34	$C_{6.5}$	1	0	0	0	0	0	0	0	0	0
35	$C_{6.6}$	1	0	0	0	0	0	0	0	0	0
36	$C_{6.7}$	1	0	0	0	0	0	0	0	0	0
37	C_7	0	0	0	0	0	0	0	0	0	0
38	$C_{7.1}$	0	0	0	0	0	0	0	0	1	0

Table A.3 The matrix IMPACT of the FR-CN of Fig. 2.11

(a)

		1	2	3	4	5	6	7	8	9	10	11	12	13	14	15
		C_1	$C_{1.1}$	$C_{1.2}$	$C_{1.3}$	$C_{1.4}$	$C_{1.5}$	$C_{1.6}$	$C_{1.7}$	C_2	$C_{2.1}$	C_3	$C_{3.1}$	$C_{3.2}$	$C_{3.2.1}$	$C_{3.3}$
1	C_1	0	0	0	0	0	0	0	0	0	0	0	0	0	0	0
2	$C_{1.1}$	0	0	0	0	0	0	0	0	0	0	0	0	0	0	0
3	$C_{1.2}$	0	0	0	0	0	0	0	0	0	0	0	0	0	0	0

(continued)

Table A.3 (continued)

(a)

		1	2	3	4	5	6	7	8	9	10	11	12	13	14	15
4	$C_{1.3}$	0	0	0	0	0	0	0	0	0	0	0	0	0	0	0
5	$C_{1.4}$	0	0	0	0	0	0	0	0	0	0	0	0	0	0	0
6	$C_{1.5}$	0	0	0	0	0	0	0	0	0	0	0	0	0	0	0
7	$C_{1.6}$	0	0	0	0	0	0	0	0	0	0	0	0	0	0	0
8	$C_{1.7}$	0	0	0	0	0	0	0	0	0	0	0	0	0	0	0
9	C_2	0	0	0	0	0	0	0	0	0	0	0	0	0	0	0
10	$C_{2.1}$	0	0	0	0	0	0	0	0	0	0	0	0	0	0	0
11	C_3	0	0	0	0	0	0	0	0	0	0	0	0	0	0	0
12	$C_{3.1}$	0	0	0	0	0	0	0	0	0	0	0	0	+0.5	0	+0.2
13	$C_{3.2}$	0	0	0	0	0	0	0	0	0	0	0	−1	0	0	+0.64
14	$C_{3.2.1}$	0	0	0	0	0	0	0	0	0	0	0	0	0	0	0
15	$C_{3.3}$	0	0	0	0	0	0	0	0	0	0	0	−1	−1	0	0
16	C_4	0	0	0	0	0	0	0	0	0	0	0	0	0	0	0
17	$C_{4.1}$	0	0	0	0	0	0	0	0	0	0	0	0	0	0	0
18	$C_{4.2}$	0	0	0	0	0	0	0	0	0	0	0	0	0	0	0
19	$C_{4.3}$	0	0	0	0	0	0	0	0	0	0	0	0	0	0	0
20	$C_{4.4}$	0	0	0	0	0	0	0	0	0	0	0	0	0	0	0
21	$C_{4.5}$	0	0	0	0	0	0	0	0	0	0	0	0	0	−0.29	0
22	C_5	0	0	0	0	0	0	0	0	0	0	0	0	0	0	0
23	$C_{5.1}$	0	0	0	0	0	0	0	0	0	0	0	0	0	0	0
24	$C_{5.2}$	0	0	0	0	0	0	0	0	0	0	0	0	0	0	0
25	$C_{5.3}$	0	0	0	0	0	0	0	0	0	0	0	0	0	0	0
26	$C_{5.4}$	0	0	0	0	0	0	0	0	0	0	0	0	0	0	0
27	$C_{5.5}$	0	0	0	0	0	0	0	0	0	0	0	0	0	−0.29	0
28	$C_{5.6}$	0	0	0	0	0	0	0	0	0	0	0	0	0	0	0
29	C_6	0	0	0	0	0	0	0	0	0	0	0	0	0	0	0
30	$C_{6.1}$	0	0	0	0	0	0	0	0	0	0	0	0	0	0	0
31	$C_{6.2}$	0	0	0	0	0	0	0	0	0	0	0	0	0	0	0
32	$C_{6.3}$	0	0	0	0	0	0	0	0	0	0	0	0	0	0	0
33	$C_{6.4}$	0	0	0	0	0	0	0	0	0	0	0	0	0	0	0
34	$C_{6.5}$	0	0	0	0	0	0	0	0	0	0	0	0	0	0	0
35	$C_{6.6}$	0	0	0	0	0	0	0	0	0	0	0	0	0	0	0
36	$C_{6.7}$	0	0	0	0	0	0	0	0	0	0	0	0	0	0	0
37	C_7	0	0	0	0	0	0	0	0	0	0	0	0	0	0	0
38	$C_{7.1}$	0	0	0	0	0	0	0	0	0	0	0	0	0	0	0

(b)

		16	17	18	19	20	21	22	23	24	25	26	27	28
		C_4	$C_{4.1}$	$C_{4.2}$	$C_{4.3}$	$C_{4.4}$	$C_{4.5}$	C_5	$C_{5.1}$	$C_{5.2}$	$C_{5.3}$	$C_{5.4}$	$C_{5.5}$	$C_{5.6}$
1	C_1	0	0	0	0	0	0	0	0	0	0	0	0	0
2	$C_{1.1}$	0	0	0	0	0	0	0	0	0	0	0	0	0
3	$C_{1.2}$	0	0	0	0	0	0	0	0	0	0	0	0	0

(continued)

Table A.3 (continued)

(b)

		16	17	18	19	20	21	22	23	24	25	26	27	28
4	$C_{1.3}$	0	0	0	0	0	0	0	0	0	0	0	0	0
5	$C_{1.4}$	0	0	0	0	0	0	0	0	0	0	0	0	0
6	$C_{1.5}$	0	0	0	0	0	0	0	0	0	0	0	0	0
7	$C_{1.6}$	0	0	0	0	0	0	0	0	0	0	0	0	0
8	$C_{1.7}$	0	0	0	0	0	0	0	0	0	0	0	0	0
9	C_2	0	0	0	0	0	0	0	0	0	0	0	0	0
10	$C_{2.1}$	0	0	0	0	0	0	0	0	0	0	0	0	0
11	C_3	0	0	0	0	0	0	0	0	0	0	0	0	0
12	$C_{3.1}$	0	0	0	0	0	0	0	0	0	0	0	0	0
13	$C_{3.2}$	0	0	0	0	0	0	0	0	0	0	0	0	0
14	$C_{3.2.1}$	0	0	0	0	0	+0.37	0	0	0	0	0	+0.37	0
15	$C_{3.3}$	0	0	0	0	0	0	0	0	0	0	0	0	0
16	C_4	0	0	0	0	0	0	0	0	0	0	0	0	0
17	$C_{4.1}$	0	0	0	0	0	0	0	0	0	0	0	0	0
18	$C_{4.2}$	0	0	0	+0.45	+0.81	0	0	0	+1	+0.45	+0.39	0	0
19	$C_{4.3}$	0	0	−0.42	0	+0.34	0	0	0	+0.42	+1	+0.41	0	0
20	$C_{4.4}$	0	0	−1	−0.45	0	0	0	0	+1	+0.45	+0.52	0	0
21	$C_{4.5}$	0	0	0	0	0	0	0	0	0	0	0	+1	0
22	C_5	0	0	0	0	0	0	0	0	0	0	0	0	0
23	$C_{5.1}$	0	0	0	0	0	0	0	0	0	0	0	0	0
24	$C_{5.2}$	0	0	−1	−0.45	−0.81	0	0	0	0	+0.45	+0.81	0	0
25	$C_{5.3}$	0	0	−0.42	−1	−0.34	0	0	0	−0.42	0	+0.41	0	0
26	$C_{5.4}$	0	0	−1	−0.45	−1	0	0	0	−1	−1	0	0	0
27	$C_{5.5}$	0	0	0	0	0	−1	0	0	0	0	0	0	0
28	$C_{5.6}$	0	0	0	0	0	0	0	0	0	0	0	0	0
29	C_6	0	0	0	0	0	0	0	0	0	0	0	0	0
30	$C_{6.1}$	0	0	0	0	0	0	0	0	0	0	0	0	0
31	$C_{6.2}$	0	0	0	0	0	0	0	0	0	0	0	0	0
32	$C_{6.3}$	0	0	0	0	0	0	0	0	0	0	0	0	0
33	$C_{6.4}$	0	0	0	0	0	0	0	0	0	0	0	0	0
34	$C_{6.5}$	0	0	0	0	0	0	0	0	0	0	0	0	0
35	$C_{6.6}$	0	0	0	0	0	0	0	0	0	0	0	0	0
36	$C_{6.7}$	0	0	0	0	0	0	0	0	0	0	0	0	0
37	C_7	0	0	0	0	0	0	0	0	0	0	0	0	0
38	$C_{7.1}$	0	0	0	0	0	0	0	0	0	0	0	0	0

(c)

	C_6	$C_{6.1}$	$C_{6.2}$	$C_{6.3}$	$C_{6.4}$	$C_{6.5}$	$C_{6.6}$	$C_{6.7}$	C_7	$C_{7.1}$
C_1	0	0	0	0	0	0	0	0	0	0
$C_{1.1}$	0	0	0	0	0	0	0	0	0	0
$C_{1.2}$	0	0	0	0	0	0	0	0	0	0

(continued)

Table A.3 (continued)

(c)

	C_6	$C_{6.1}$	$C_{6.2}$	$C_{6.3}$	$C_{6.4}$	$C_{6.5}$	$C_{6.6}$	$C_{6.7}$	C_7	$C_{7.1}$
$C_{1.3}$	0	0	0	0	0	0	0	0	0	0
$C_{1.4}$	0	0	0	0	0	0	0	0	0	0
$C_{1.5}$	0	0	0	0	0	0	0	0	0	0
$C_{1.6}$	0	0	0	0	0	0	0	0	0	0
$C_{1.7}$	0	0	0	0	0	0	0	0	0	0
C_2	0	0	0	0	0	0	0	0	0	0
$C_{2.1}$	0	0	0	0	0	0	0	0	0	0
C_3	0	0	0	0	0	0	0	0	0	0
$C_{3.1}$	0	0	0	0	0	0	0	0	0	0
$C_{3.2}$	0	0	0	0	0	0	0	0	0	0
$C_{3.2.1}$	0	0	0	0	0	0	0	0	0	0
$C_{3.3}$	0	0	0	0	0	0	0	0	0	0
C_4	0	0	0	0	0	0	0	0	0	0
$C_{4.1}$	0	0	0	0	0	0	0	0	0	0
$C_{4.2}$	0	0	0	0	0	0	0	0	0	0
$C_{4.3}$	0	0	0	0	0	0	0	0	0	0
$C_{4.4}$	0	0	0	0	0	0	0	0	0	0
$C_{4.5}$	0	0	0	0	0	0	0	0	0	0
C_5	0	0	0	0	0	0	0	0	0	0
$C_{5.1}$	0	0	0	0	0	0	0	0	0	0
$C_{5.2}$	0	0	0	0	0	0	0	0	0	0
$C_{5.3}$	0	0	0	0	0	0	0	0	0	0
$C_{5.4}$	0	0	0	0	0	0	0	0	0	0
$C_{5.5}$	0	0	0	0	0	0	0	0	0	0
$C_{5.6}$	0	0	0	0	0	0	0	0	0	0
C_6	0	0	0	0	0	0	0	0	0	0
$C_{6.1}$	0	0	0	0	+0.43	0	0	0	0	0
$C_{6.2}$	0	0	0	0	0	0	0	0	0	0
$C_{6.3}$	0	0	0	0	0	0	0	0	0	0
$C_{6.4}$	0	−0.51	0	0	0	+0.33	+0.33	+0.27	0	0
$C_{6.5}$	0	0	0	0	−0.99	0	+0.77	0	0	0
$C_{6.6}$	0	0	0	0	−0.99	−0.77	0	0	0	0
$C_{6.7}$	1	0	0	0	−0.78	0	0	0	0	0
C_7	0	0	0	0	0	0	0	0	0	0
$C_{7.1}$	0	0	0	0	0	0	0	0	1	0

Appendix B: Questionnaires

Questionnaire A

Table B.1 Questionnaire for measuring learners' satisfaction

The questionnaire					
Questions	Answers (circle one for each question)				
1 Was the content well organized?	1 (not at all)	2 (slightly)	3 (moderately)	4 (very)	5 (absolutely)
2 Did the presentation of the content meet your needs?	1 (not at all)	2 (slightly)	3 (moderately)	4 (very)	5 (absolutely)
3 Was the content practical and useful?	1 (not at all)	2 (slightly)	3 (moderately)	4 (very)	5 (absolutely)
4 Were exercises and tests useful?	1 (not at all)	2 (slightly)	3 (moderately)	4 (very)	5 (absolutely)
5 Did the educational system keep your interest alive?	1 (not at all)	2 (slightly)	3 (moderately)	4 (very)	5 (absolutely)
6 What is your opinion about the quality of instruction?	1 (very poor)	2 (poor)	3 (fair)	4 (good)	5 (excellent)
7 Did you feel that you had assimilated all the subjects that you are taught?	1 (not at all)	2 (slightly)	3 (moderately)	4 (very)	5 (absolutely)
8 Was the user interface friendly?	1 (not at all)	2 (slightly)	3 (moderately)	4 (very)	5 (absolutely)
9 Did the educational system meet your expectations?	1 (not at all)	2 (slightly)	3 (moderately)	4 (very)	5 (absolutely)

(continued)

© Springer International Publishing Switzerland 2015
K. Chrysafiadi and M. Virvou, *Advances in Personalized Web-Based Education*,
Intelligent Systems Reference Library 78, DOI 10.1007/978-3-319-12895-5

Table B.1 (continued)

The questionnaire

Questions	Answers (circle one for each question)				
10 Is the educational system useful as an educational tool for programming languages?	1 (not at all)	2 (slightly)	3 (moderately)	4 (very)	5 (absolutely)
11 Do you think that the use of the particular educational system was waste of time?	1 (not at all)	2 (slightly)	3 (moderately)	4 (very)	5 (absolutely)
12 Did the educational system correspond to your needs and knowledge level each time?	1 (not at all)	2 (slightly)	3 (moderately)	4 (very)	5 (absolutely)
13 How time did you spend on issues that you already known?	1 (not at all)	2 (slightly)	3 (moderately)	4 (very)	5 (very much)
14 Did the prompt for revision was useful and appropriate?	1 (not at all)	2 (slightly)	3 (moderately)	4 (very)	5 (absolutely)
15 What is your overall rating?	1 (very poor)	2 (poor)	3 (fair)	4 (good)	5 (excellent)

The answers

	Questions		Evaluation degree
Quality of content	1	Was the content well organized?	4.44
	2	Did the presentation of the content meet your needs?	4.1
	3	Was the content practical and useful?	4.62
	4	Were exercises and tests useful?	4.62
Quality of instruction	5	Did the educational system keep your interest alive?	4.1
	6	What is your opinion about the quality of instruction?	4.18
	7	Did you feel that you had assimilated all the subjects that you are taught?	4.44
Friendliness	8	Was the user interface friendly?	3.97
Usefulness	9	Did the educational system meet your expectations?	4
	10	Is the educational system useful as an educational tool for programming languages?	4.24
	11	Do you think that the use of the particular educational system was waste of time?	1

(continued)

Table B.1 (continued)

The answers			
	Questions		Evaluation degree
Adaptivity	12	Did the educational system correspond to your needs and knowledge level each time?	4.15
	13	How time did you spend on issues that you already known?	1.73
	14	Did the prompt for revision was -useful and appropriate?	4.06
Overall rating	15	What is your overall rating?	4.38

Questionnaire B

Table B.2 The questionnaire for measuring the changes on learners' behavior and thoughts about computer programming

The questionnaire					
Questions	Answers (circle one for each question)				
1 What is you opinion about computer programming?	1 (not at all interesting)	2 (slightly interesting)	3 (moderately interesting)	4 (very interesting)	5 (extremely interesting)
2 Do you want to learn a (another) programming language?	1 (not at all)	2 (slightly)	3 (moderately)	4 (very)	5 (absolutely)
3 Do you interested in participating in a computer-programming project?	1 (not at all)	2 (slightly)	3 (moderately)	4 (very)	5 (absolutely)
4 Are you motivated to use computer programming in your job?	1 (not at all)	2 (slightly)	3 (moderately)	4 (very)	5 (absolutely)
5 Do you think that computer programming can facilitate some everyday processes?	1 (strongly disagree)	2 (disagree)	3 (neither agree or disagree)	4 (agree)	5 (strongly agree)

(continued)

Table B.2 (continued)

The answers						
Questions	Evaluation degree					
	Before			After		
	Arts	Science (other than computers)	Computer science related	Arts	Science (other than computers)	Computer science related
1 What is you opinion about computer programming?	2.7	3.6	4	3.92	4.2	4.42
2 Do you want to learn a (another) programming language?	3.54	4.2	4.25	3.69	4.5	4.58
3 Do you interested in participating in a computer-programming project?	2.23	3.3	4.1	3.46	4.1	4.5
4 Are you motivated to use computer programming in your job?	1.77	3.3	4.34	3.39	4.4	4.83
5 Do you think that computer programming can facilitate some everyday processes?	2.69	3.9	4.58	4.08	4.5	4.67

Questionnaire C

Table B.3 The questionnaire for measuring the changes on learners' behavior and thoughts about e-learning

The questionnaire					
Questions	Answers (circle one for each question)				
1 What is you opinion about distance learning?	1 (not at all interesting)	2 (slightly interesting)	3 (moderately interesting)	4 (very interesting)	5 (extremely interesting)
2 How useful do you think that is the computer-based education?	1 (not at all)	2 (slightly)	3 (moderately)	4 (very)	5 (absolutely)

(continued)

Table B.3 (continued)

The questionnaire

Questions	Answers (circle one for each question)				
3 Do you interested in participating in an e-learning training program?	1 (not at all)	2 (slightly)	3 (moderately)	4 (very)	5 (absolutely)
4 Do you think that e-learning systems can provide an effective education?	1 (strongly disagree)	2 (disagree)	3 (neither agree or disagree)	4 (agree)	5 (strongly agree)
5 Do you think that e-learning systems can facilitate the educational process?	1 (strongly disagree)	2 (disagree)	3 (neither agree or disagree)	4 (agree)	5 (strongly agree)

The answers

Questions	Evaluation degree					
	Before			After		
	Arts	Science (other than computers)	Computer science related	Arts	Science (other than computers)	Computer science related
1 What is you opinion about distance learning?	3	3.4	3.75	3.8	3.8	4
2 How useful do you think that is the computer-based education?	2.85	3.4	3.67	3.77	3.6	4
3 Do you interested in participating in an e-learning training program?	3.39	3.7	4.08	3.85	4.1	4.42
4 Do you think that e-learning systems can provide an effective education?	2.46	3.2	3.67	4.08	4.3	4.08
5 Do you think that e-learning systems can facilitate the educational process?	2.46	3	3.92	4.23	4.1	4.25

Questionnaire D

Table B.4 The questionnaire for measuring the system's results on learners' further studies

The questionnaire					
Questions	Answers (circle one for each question)				
1 Did the educational software help you to understand better the logic of programming?	1 (not at all)	2 (slightly)	3 (moderately)	4 (very)	5 (absolutely)
2 Did the educational software help you to learn other programming languages?	1 (not at all)	2 (slightly)	3 (moderately)	4 (very)	5 (absolutely)
3 Did the educational software help you in your studies?	1 (not at all)	2 (slightly)	3 (moderately)	4 (very)	5 (absolutely)
4 Did the educational software help you to understand better other lessons of computer science?	1 (not at all)	2 (slightly)	3 (moderately)	4 (very)	5 (absolutely)
5 Did the educational software help you in the elaboration of tasks and activities considering your studies?	1 (not at all)	2 (slightly)	3 (moderately)	4 (very)	5 (absolutely)

The answers	
Questions	Evaluation degree
1 Did the educational software help you to understand better the logic of programming?	4.4
2 Did the educational software help you to learn other programming languages?	3.86
3 Did the educational software help you in your studies?	3.95
4 Did the educational software help you to understand better other lessons of computer science?	3.34
5 Did the educational software help you in the elaboration of tasks and activities considering your studies?	3.71

Questionnaire E

Table B.5 The questionnaire for measuring learners' satisfaction about the system's adaptive responses to their needs

The questionnaire					
Questions	Answers (circle one for each question)				
1	Did the educational system correspond to your needs and knowledge level each time?	1 (not at all)	2 (slightly)	3 (moderately)	4 (very)	5 (absolutely)
2	How time did you spend on issues that you already known?	1 (not at all)	2 (slightly)	3 (moderately)	4 (very)	5 (absolutely)
3	Did the prompt for revision was useful and appropriate?	1 (not at all)	2 (slightly)	3 (moderately)	4 (very)	5 (absolutely)
4	Did you need to read some concepts that the system considered to be learned?	1 (not at all)	2 (slightly)	3 (moderately)	4 (very)	5 (absolutely)
5	Did the system return you to read again a concept that you knew it?	1 (not at all)	2 (slightly)	3 (moderately)	4 (very)	5 (absolutely)
6	Were the returns to already learned concepts meaningful?	1 (not at all)	2 (slightly)	3 (moderately)	4 (very)	5 (absolutely)
7	Was the system's decision not to read some concepts meaningful?	1 (not at all)	2 (slightly)	3 (moderately)	4 (very)	5 (absolutely)
8	Did the system's inferences about your knowledge level on each domain concept correspond to your actual needs and level?	1 (not at all)	2 (slightly)	3 (moderately)	4 (very)	5 (absolutely)

The answers		
Questions		Evaluation degree
1	Did the educational system correspond to your needs and knowledge level each time?	4.15
2	How time did you spend on issues that you already known?	1.73
3	Did the prompt for revision was useful and appropriate?	4.06
4	Did you need to read some concepts that the system considered to be learned?	1.44
5	Did the system return you to read again a concept that you knew it?	1.94
6	Were the returns to already learned concepts meaningful?	4.03
7	Was the system's decision not to read some concepts meaningful?	4.56
8	Did the system's inferences about your knowledge level on each domain concept correspond to your actual needs and level?	4.08

Questionnaire F

Table B.6 The questionnaire for measuring the validity of the adaptation decision making

The questionnaire					
Questions	Answers (circle one for each question)				
1	Were the returns to a previous learned concept for revision a waste of time?	1 (not at all)	2 (slightly)	3 (moderately)	4 (very)	5 (absolutely)
2	How many times the returns to a previous read concept for revision concerned concepts that you actually knew?	1 (not at all)	2 (slightly)	3 (moderately)	4 (very)	5 (absolutely)
3	Did the prompt for revision was useful and appropriate?	1 (not at all)	2 (slightly)	3 (moderately)	4 (very)	5 (absolutely)
4	Did the returns to a previous domain concept correspond to your need for revision?	1 (not at all)	2 (slightly)	3 (moderately)	4 (very)	5 (absolutely)
5	Were the returns to already learned concepts meaningful?	1 (not at all)	2 (slightly)	3 (moderately)	4 (very)	5 (absolutely)
6	Did the returns to a previous domain concept for revision help you to learn computer programming better?	1 (not at all)	2 (slightly)	3 (moderately)	4 (very)	5 (absolutely)

The answers		
Questions		Evaluation degree
1	Were the returns to a previous learned concept for revision a waste of time?	1.73
2	How many times the returns to a previous read concept for revision concerned concepts that you actually knew?	2.06
3	Did the prompt for revision was useful and appropriate?	4.06
4	Did the returns to a previous domain concept correspond to your need for revision?	3.86
5	Were the returns to already learned concepts meaningful?	4.03
6	Did the returns to a previous domain concept for revision help you to learn computer programming better?	4

Appendix C: Screenshots

- Registration and log-in (Figs. C.1 and C.2).
- Domain concepts (Figs. C.3, C.4 and C.5).
- Exercises and questions of tests (Figs. C.6, C.7, C.8 and C.9).
- Results of the tests (Figs. C.10 and C.11).

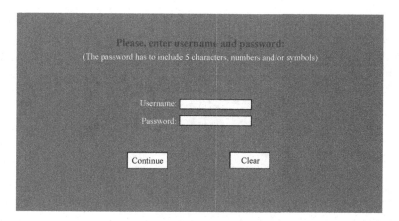

Fig. C.1 Log-in form

© Springer International Publishing Switzerland 2015
K. Chrysafiadi and M. Virvou, *Advances in Personalized Web-Based Education*,
Intelligent Systems Reference Library 78, DOI 10.1007/978-3-319-12895-5

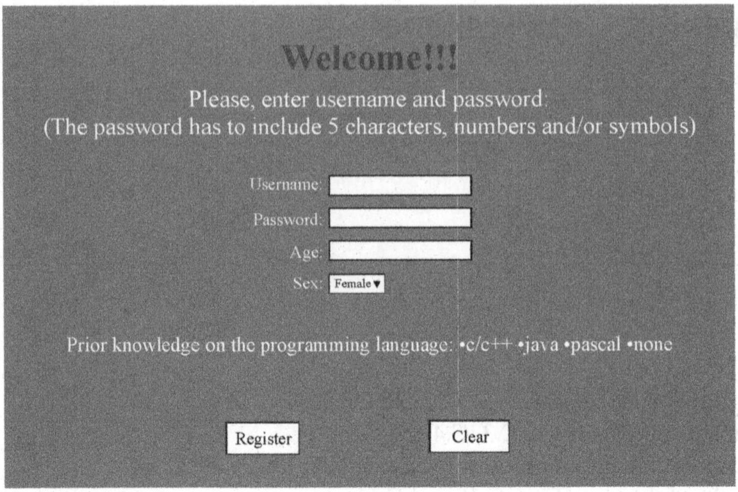

Welcome!!!

Please, enter username and password:
(The password has to include 5 characters, numbers and/or symbols)

Username:

Password:

Age:

Sex: Female ▼

Prior knowledge on the programming language: •c/c++ •java •pascal •none

Register   Clear

Fig. C.2 Registration form

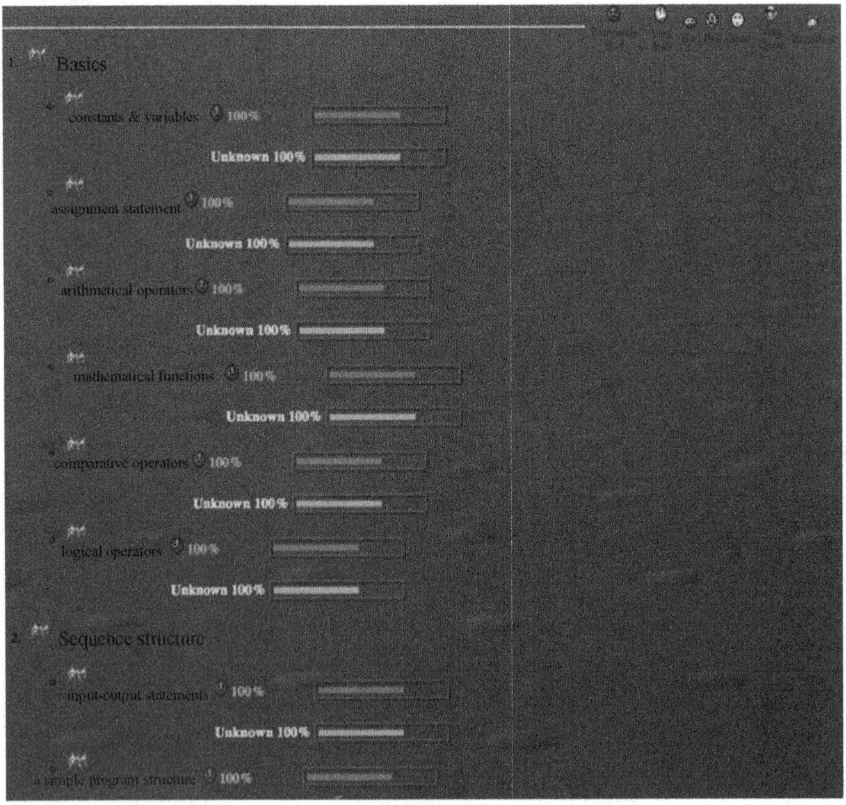

1. Basics

   constants & variables 100%

   Unknown 100%

   assignment statement 100%

   Unknown 100%

   arithmetical operators 100%

   Unknown 100%

   mathematical functions 100%

   Unknown 100%

   comparative operators 100%

   Unknown 100%

   logical operators 100%

   Unknown 100%

2. Sequence structure

   input-output statements 100%

   Unknown 100%

   a simple program structure 100%

Fig. C.3 Domain concepts of knowledge stereotype 1

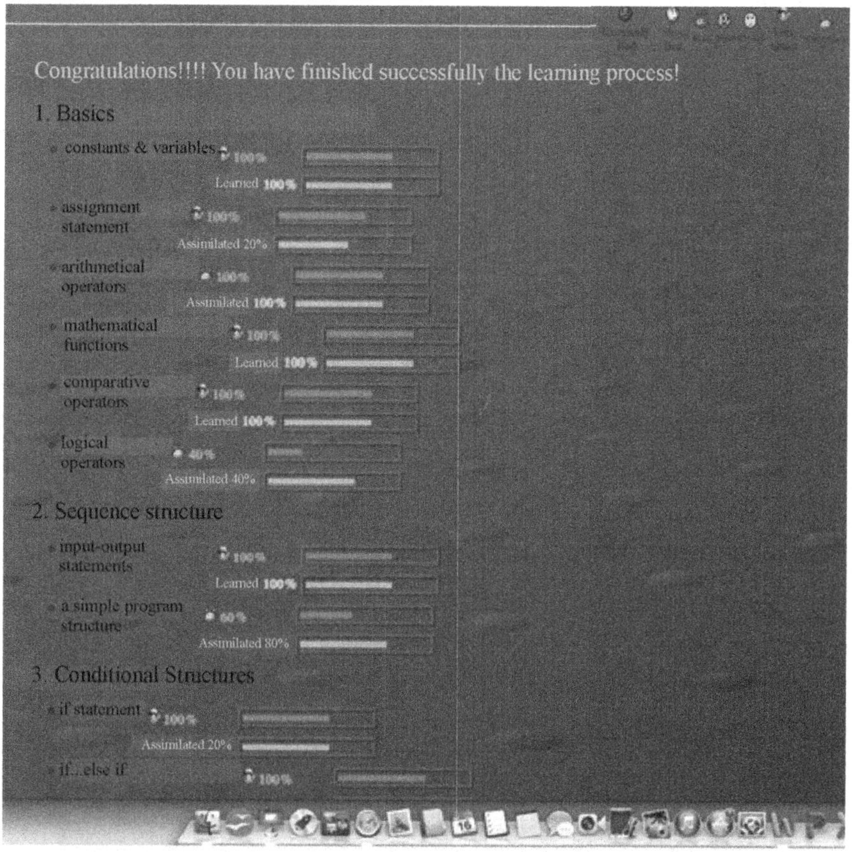

Fig. C.4 Successful completion of the learning process of all the learning material

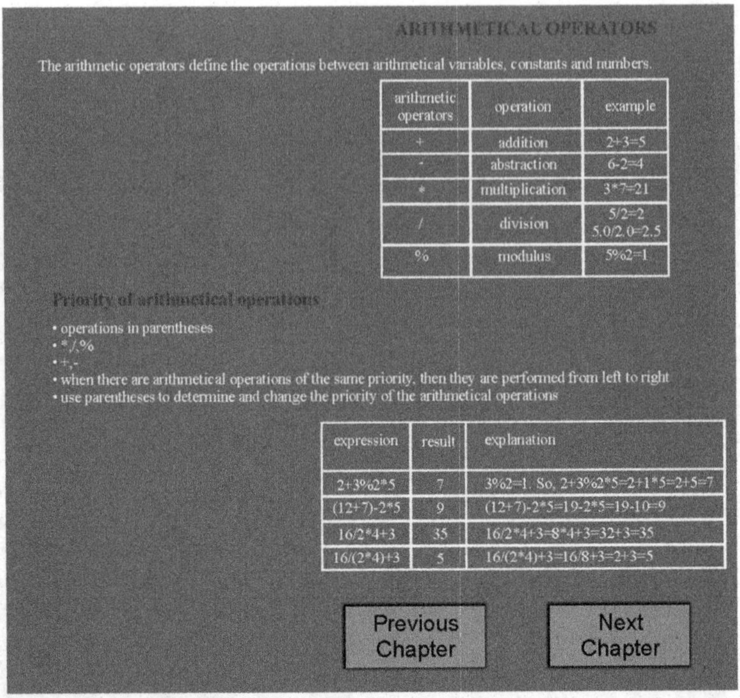

Fig. C.5 Arithmetic operators

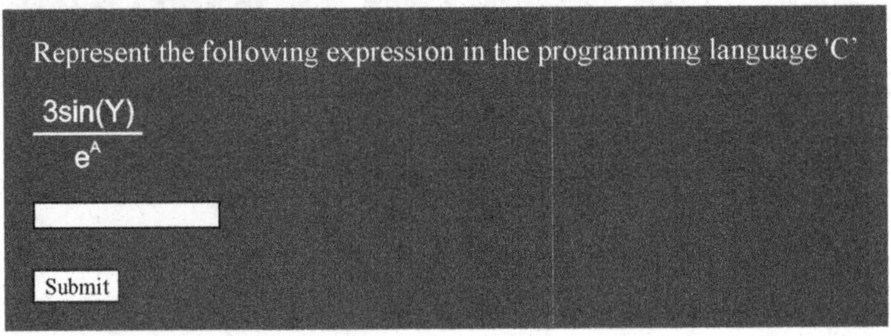

Fig. C.6 Fill in the gaps exercise

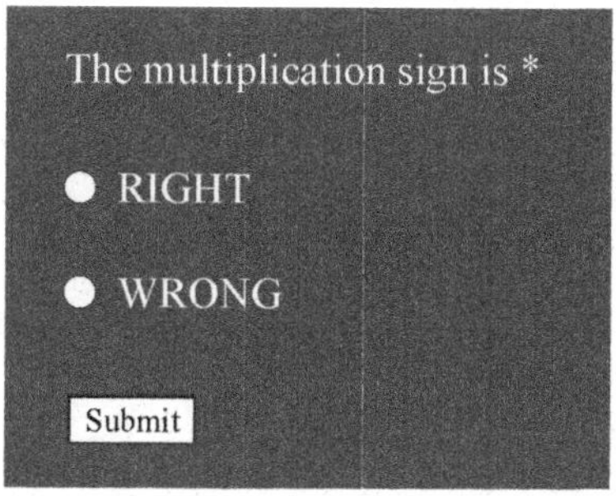

The multiplication sign is *

● RIGHT

● WRONG

Submit

Fig. C.7   Right-wrong exercise

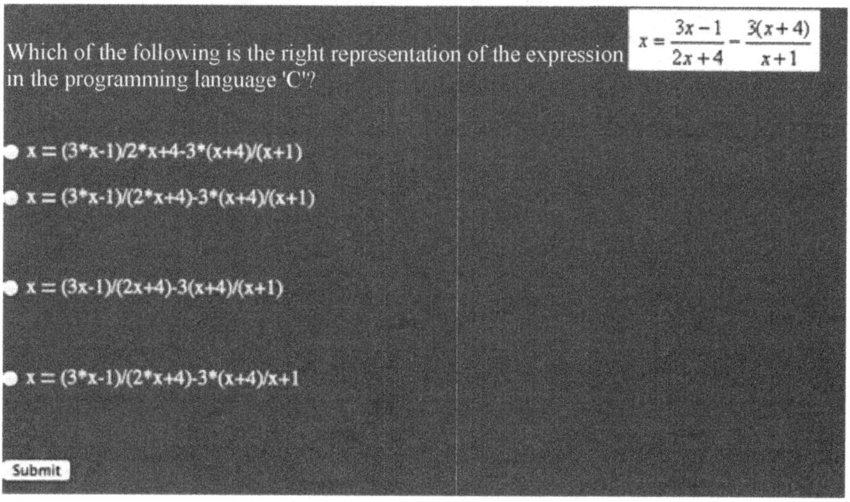

Which of the following is the right representation of the expression $x = \dfrac{3x-1}{2x+4} - \dfrac{3(x+4)}{x+1}$ in the programming language 'C'?

● x = (3*x-1)/2*x+4-3*(x+4)/(x+1)

● x = (3*x-1)/(2*x+4)-3*(x+4)/(x+1)

● x = (3x-1)/(2x+4)-3(x+4)/(x+1)

● x = (3*x-1)/(2*x+4)-3*(x+4)/x+1

Submit

Fig. C.8   Multiple-choice exercise

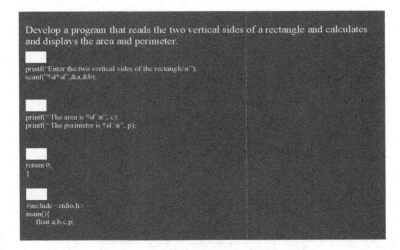

Fig. C.9 Put the algorithm pieces in the right order exercise

test's id	l e v e l	constants & variables	assignment statement	arithmetical operators	mathematical functions	comparative operators	logical operators	TOTAL ERRORS DUE TO PRIOR KNOWL EDGE	TOTAL SYNTAX ERRORS	TOTAL LOGICAL ERRORS
416	1	PRIOR KNOWLEDGE= 0% SYNTAX=60% LOGICAL=40%	PRIOR KNOWLEDGE =50% SYNTAX =100% LOGICA=0%	PRIOR KNOWLEDGE =0% SYNTAX = 0% LOGICAL=100%	PRIOR KNOWLEDGE =0% SYNTAX = 0% LOGICAL=0%	PRIOR KNOWLEDGE =0% SYNTAX =0% LOGICAL=0%	PRIOR KNOWLEDGE =0% SYNTAX = 0% LOGICAL=0%	12%	62%	37%
417	1	PRIOR KNOWLEDGE=0 0% SYNTAX =100% LOGICA=0%	PRIOR KNOWLEDGE =0% SYNTAX = 0% LOGICAL=0%	PRIOR KNOWLEDGE=0 0% SYNTAX =100% LOGICA=0%	PRIOR KNOWLEDGE =0% SYNTAX = 0% LOGICAL=0%	PRIOR KNOWLEDGE =0% SYNTAX =0% LOGICAL=0%	PRIOR KNOWLEDGE =0% SYNTAX = 0% LOGICAL=0%	0%	100%	0%

Fig. C.10 Overall results-progress

QUESTION	ANSWER	NOTICE	
The assignment statement A = B changes the value of the variable B.	RIGHT		CHECK
A logical variable can take 3 values.	WRONG		CHECK
We can use only constants for representing data in an algorithm.	WRONG		CHECK
A constant does not change during execution of algorithms.	WRONG		CHECK
Variables take values that may be characters.	RIGHT		CHECK
Which of the following instructions assigns the value 138 to the variable A?	WRONG	Your answer is right for the programming language PASCAL	CHECK
What is the value of the variable Π after executing the following command P=(3+4/2*3)*2-(3*2+4-2)*2+9/3+40?	WRONG		CHECK
What the following program segment will display on the screen? A = 'B'; B = 'A'; C = A; Printf("B %c %c", A,C);	RIGHT		CHECK
What type of variable we will use to represent an employee's name?	WRONG		CHECK
What type of variable we will use to represent the number of students in a class?	RIGHT		CHECK

Read again the following concepts:

Constants & variables

Assignment statement

Arithmetical operators

Fig. C.11 Results of the test

References

Al-Hmouz, A., Shen, J., Yan, J., & Al-Hmouz, R. (2010). Enhanced learner model for adaptive mobile learning. In *Proceedings of the 12th International Conference on Information Integration and Web-based Applications and Services,* Paris, France (pp. 783–786).

Al-Hmouz, A., Shen, J., Yan, J., & Al-Hmouz, R. (2011). Modeling mobile learning system using ANFIS. In *Proceedings of the 11th IEEE International Conference on Advanced Learning Technologies (ICALT 2011),* Athens, USA (pp. 32–36).

Albano, G. (2011). Knowledge, skills, competencies: A model for mathematics e-learning. In *Proceedings of the 18th International Conference on Telecommunications,* Ayia Napa, Cyprus (pp. 214–225).

Alepis, E., & Virvou, M. (2006). Emotional intelligence: constructing user stereotypes for affective bi-modal interaction. In Proc. of the 10th International Conference on Knowledge-Based & Intelligent Informationa & Engineering Systems (KES2006), Bournemouth International Conference Centre, United Kingdom (pp. 435–442).

Alepis, E., & Virvou, M. (2011). Automatic generation of emotions in tutoring agents for affective e-learning in medical education. *Expert Systems with Applications, 38,* 9840–9847.

Alepis, E., Virvou, M., & Kabassi., K. (2008). Mobile education: Towards affective bi-modal interaction for adaptivity. In *Proceedings of the 3rd International Conference on Digital Information Management, ICDIM 2008,* London (pp. 51–56).

Alfonseca, E., Carro, R.M., Martin, E., Ortigosa, A., & Paredes, P. (2006). The impact of learning styles on student grouping for collaborative learning: a case study. *User Modeling and User-Adapted Interaction, 16*(3–4), 377–401.

Almohammadi, K., & Hagras, H. (2013). An adaptive fuzzy logic based system for improved knowledge delivery within intelligent e-learning platforms. In *Proceedings of the b2013 International Conference on Fuzzy Systems* (pp. 1–8). India: Hyderabad International Convention Center (HICC).

Alves, P., Amaral, L., & Pires, J. (2008). Case-based reasoning approach to adaptive web-based educational systems. In *Proceedings Eighth IEEE International Conference on Advanced Learning Technologies (ICALT '08),* Santander, Cantabria, Spain (pp. 260–261).

Arroyo, I., Meheranian, H., & Woolf, B. P. (2010). Effort-based tutoring: An empirical approach to intelligent tutoring. In *Proceedings of the 3rd International Conference on Educational Data Mining* (pp.1–10). Pittsburgh, PA, USA.

Azadeh, A., Ziaei, B., & Moghaddam, M. (2012). A hybrid fuzzy regression-fuzzy cognitive map algorithm for forecasting and optimization of housing market fluctuations. *Expert Systems with Applications, 39*(1), 298–315.

Badaracco, M., & Martinez, L. (2013). A fuzzy linguistic algorithm for adaptive test in intelligent tutoring system based on competences. *Expert Systems with Applications, 40*(8), 3073–3086.

© Springer International Publishing Switzerland 2015

K. Chrysafiadi and M. Virvou, *Advances in Personalized Web-Based Education,*

Intelligent Systems Reference Library 78, DOI 10.1007/978-3-319-12895-5

Baghaei, N., Mitrovic, A., & Irwin, W. (2005). A constraint-based tutor for learning Object-Oriented analysis and design using UML. *Proceedings. International Conference on Computers in Education 2005*, Singapore (pp. 9–16).

Bai, S. M., & Chen, S. M. (2006a). Automatically constructing grade membership functions for students' evaluation for fuzzy grading systems. In *Proceedings of the 2006 World Automation Congress*. Budapest, Hungary.

Bai, S. M., & Chen, S. M. (2006b). A new method for students' learning achievement using fuzzy membership functions. In *Proceedings of the 11th Conference on Artificial Intelligence, Kaohsiung, Taiwan, Republic of China*.

Bai, S. M., & Chen, S. M. (2008). Evaluating students' learning achievement using fuzzy membership functions and fuzzy rules. *Expert Systems with Applications, 34*, 399–410.

Baker, R. S. (2007). Modeling and understanding students' off-task behavior in intelligent tutoring systems. In *Proceedings of the SIGCHI Conf. on Human Factors in Computing Systems*, San Jose, California, USA (pp. 1059–1068).

Baker, R. S., Corbett, A. T., Koedinger, K. R., & Wagner, A. Z. (2004). Off-task behavior in the cognitive tutor classroom: When students 'game the system. In *Proceedings of the SIGCHI Conference on Human Factors in Computing Systems (CHI '04)*, Vienna, Austria (pp. 383–390).

Baker, R. S., Goldstein, A. B., & Heffernan, N. T. (2010). Detecting the moment of learning. In *Proceedings of the ACM International Conference on Interactive Tabletops and Surfaces*, Saarbrücken, Germany (pp. 25–34).

Barak, M. (2010). Motivating self-regulated learning in technology education. *International Journal of Technology and Design Education, 20*(4), 381–401.

Balakrishnan, A. (2011). On Modeling the affective effect on learning. In *Proc. of the 5th international conference on Multi-Disciplinary Trends in Artificial Intelligence*, Hyderabald, India (pp. 225–235).

Baschera, G. M., & Gross, M. (2010). Poisson-based inference for perturbation models in adaptive spelling training. *International Journal of Artificial Intelligence in Education, 20*, 1–3.

Bishop, C. C., & Wheeler, D. (1994). The myers-briggs personality type and its relationship to computer programming. *Journal of Research on Computing in Education, 26*, 358–371.

Biswas, R. (1995). An application of fuzzy sets in students' evaluation. *Fuzzy Sets and Systems, 74*(2), 187–194.

Bloom, B. S. (1984). The 2 sigma problem: The search for methods of group instruction as effective as one-to-one tutoring. *Educational Researcher, 13*(6), 4–16.

Bontcheva, K., & Wilks, Y. (2005). Tailoring automatically generated hypertext. *User Modeling and User-Adapted Interaction, 15*(1–2), 135–168.

Boticario, J. G., Santos, O. C., & van Rosmalen, P. (2005). Issues in developing standard-based adaptive learning management systems. In *Proceedings of the EADTU 2005 Working Conference: Towards Lisbon 2010: Collaboration for Innovative Content in Lifelong Open and Flexible Learning*.

Brusilovsky, P., & Anderson, J. (1998). ACT-R electronic bookshelf: An adaptive system for learning cognitive psychology on the Web. In *Proceedings of the WebNet'98, World Conference of the WWW, Internet, and Intranet*, Orlando, Florida, USA (pp. 92–97).

Brusilovsky, P., & Millán, E. (2007). User models for adaptive hypermedia and adaptive educational systems. In P. Brusilovsky, A. Kobsa, & W. Neidl (Eds.), *The Adaptive Web: Methods and Strategies of Web Personalization* (pp. 3–53).

Bunt, A., & Conati, C. (2003). Probabilistic student modelling to improve exploratory behaviour. *User Modeling and User-Adapted Interaction, 13*(3), 269–309.

Burstein, M. H., & Collins, A. M. (1988). Modeling a theory of human plausible reasoning. In T. O'Shea & V. Sgurev (Eds.), *Artificial intelligence iii: methodology, systems, applications* (pp. 21–28). North Holland: Elsevier Science Publishers B.V.

Burstein, M. H., Collins, A., & Baker, M. (1991). Plausible generalisation: extending a model of human plausible reasoning. *Journal of the Learning Sciences, 3*, 319–359.

Carmona, C., & Conejo, R. (2004). A learner model in a distributed environment. In *Proceedings of the 3rd International Conference on Adaptive Hypermedia and Adaptive Web-Based Systems (AH'2004)*, Eindhoven, the Netherlands (pp. 353–359).

Carmona, C., Castillo, G., & Millán, E. (2008). Designing a dynamic bayesian network for modeling students' learning styles. In *Proceedings of the 8th IEEE International Conference on Advanced Learning Technologies (ICALT 2008)*, Santander, Cantabria, Spain (pp. 346–350).

Carver, R., & Nash, J. G. (2009). *Doing data analysis with SPSS*. United States: Cengage Learning Inc.

Castillo, O., & Melin, P. (2008). Intelligent systems with interval type-2 fuzzy logic. *International Journal of Innovative Computing, Information and Control, 4*(4), 771–783.

Castillo, G., Gama, J., & Breda, A. M. (2009). An adaptive predictive model for student modeling. In *Advances in Web-Based Education: Personalized Learning Environments* (pp. 70–92). USA: Information Science Publishing (Chapter IV).

Cetintas, S., Si, L., Xin, Y. P., & Hord C. (2010). Automatic detection of off-task behaviors in intelligent tutoring systems with machine learning techniques. *IEEE Transactions on Learning Technologies, 3*(3), 228–236.

Chang, D. F., & Sun, C. M. (1993). Fuzzy assessment of learning performance of junior high school students. In *Proceedings of the 1993 first national symposium on fuzzy theory and applications*, Hsinchu, Taiwan, Republic of China (pp. 1–10).

Chen, S. M., & Lee, C. H. (1999). New methods for students' evaluating using fuzzy sets. *Fuzzy Sets and Systems, 104*(2), 209–218.

Cheung, R., Wan, C., & Cheng, C. (2010). An ontology-based framework for personalized adaptive learning. In *Proceedings of the 9th International Conference on Web-based Learning (ICWL 2010)* (pp. 52–61). Shanghai, China.

Chieu, V. M., Luengo, V., Vadcard, L., & Tonetti, J. (2010). Student modeling in orthopedic surgery training: Exploiting symbiosis between temporal bayesian networks and fine-grained didactic analysis. *Journal of Artificial Intelligence in Education, 20*(3), 269–301.

Chin, D. N. (2001). Empirical evaluation of the user models and user-adapted systems. *User Modelling and User-Adapted Interaction, 11*, 181–194.

Cho, M.-H., & Kin, B. J. (2013). Students' self-regulation for interaction with others in online learning environments. *Internet and Higher Education, 17*, 69–75.

Chrysafiadi, K., & Virvou, M. (2008). Personalized teaching of a programming language over the web: Stereotypes and rule-based mechanisms. In *Proceedings of the 8th Joint Conference on Knowledge-Based Software Engineering*, Piraeus, Greece (pp. 484–492).

Chrysafiadi, K., & Virvou, M. (2012). Evaluating the integration of fuzzy logic into the student model of a web-based learning environment. *Experts Systems with Applications, 39*(18), 13127–13134.

Chrysafiadi, K., & Virvou, M. (2013a). PeRSIVA: An empirical evaluation method of a student model of an intelligent e-learning environment for computer programming. *Computers and Education, 68*, 322–333.

Chrysafiadi, K., & Virvou, M. (2013b). Student modeling approaches: A literature review for the last decade. *Expert Systems with Applications, 40*(11), 4715–4729.

Chrysafiadi, K., & Virvou, M. (2013c). Dynamically personalized e-training in computer programming and the language C. *IEEE Transactions on Education, 56*(4), 385–392.

Chrysafiadi, K., & Virvou, M. (2014). Fuzzy Logic for adaptive instruction in an e-learning environment for computer programming. *IEEE Transactions on Fuzzy Systems*. d.o.i.:10.1109/TFU22.2014.2310242.

Clancey, W. (1988). The role of qualitative models in instruction. In J. Self (eds.), Artificial Intelligence and Human Learning, Chapman and Hall Computing.

Clemente, J., Ramírez, J., & de Antonio, A. (2011). A proposal for student modeling based on ontologies and diagnosis rules. *Expert Systems with Applications, 38*(7), 8066–8078.

Codara, L. (1998). *Le mappe cognitive*. Roma: Carrocci Editore.

Collins, A., & Michalski, R. (1989). The logic of plausible reasoning: A core theory. *Cognitive science* (Vol. 13, pp. 1–49). The Netherlands: Elsevier Science.

Conati, C. (2009). Intelligent tutoring systems: New challenges and directions. In *Proceedings of the 21st International Joint Conference on Artificial Intelligence* (pp. 2–7). San Francisco, CA, USA.

Conati, C., & Maclaren, H. (2009). Empirically building and evaluating a probabilistic model of user affect. *User Modeling and User-Adapted Interaction, 19*(3), 267–303.

Conati, C., & Zhou, X. (2002). Modeling students' emotions from cognitive appraisal in educational games. In *Proceedings of the 6th International Conference on Intelligent Tutoring Systems*, Biarritz, France and San Sebastian, Spain (pp. 944–954).

Conati, C., Gertner, A., & Vanlehn, K. (2002). Using bayesian networks to manage uncertainty in student modeling. *User Modeling and User-Adapted Interaction, 12*(4), 371–417.

Craiger, J. P., Goodman, D. F., Weiss, R. J., & Butler, A. (1996). Modeling organizational behavior with fuzzy cognitive maps. *Journal of Computational Intelligence and Organizations, 1*, 120–123.

Crockett, K., Latham, A., McLean, D., & O'Shea, J. (2013). A fuzzy model for predicting learning styles using behavioral cues in an conversational intelligent tutoring system. In *Proceedings of the b2013 International Conference on Fuzzy Systems* (pp. 1–8). India: Hyderabad International Convention Center (HICC).

de Raadt, M. (2007). A review of Australasian investigations into problem solving and the novice programmer. *Computer Science Education, 17*(3), 201–213.

Del Missier, F., & Ricci, F. (2003). Understanding recommender systems: Experimental evaluation challenges. In *Proceedings of the Second Workshop on Empirical Evaluation of Adaptive Systems*, Pittsburgh (pp. 31–40).

Dempster, J. (2004). Evaluating e-learning developments: An overview. Retrieved July 23, 2008 from http://www.warwick.ac.uk/go/cap/resources/eguides.

Desmarais, M. C., & Baker, R. S. (2012). A review of recent advances in learner and skill modeling in intelligent learning environments. *User Modeling and User-Adapted Interaction, 22*(1–2), 9–38.

Devedzic, V. (2006). *Semantic web and education (Monograph)*. Berlin Heidelberg, New York: Springer.

Dodds, P., & Fletcher, J. (2004). Opportunities for new "smart" learning environments enabled by next-generation web capabilities. *Journal of Educational Multimedia and Hypermedia, 13*(4), 391–404.

Dolog, P., Henze, N., Nejdl, W., & Sintek, M. (2004). The personal reader: Personalizing and enriching learning resources using semantic web technologies. In *Proceedings of the 3rd International Conference on Adaptive Hypermedia and Adaptive Web-Based Systems*, Eindhoven, Netherlands (pp. 85–94).

Drigas, A., Argyri, K., & Vrettaros, J. (2009). Decade review (1999–2009): Artificial intelligence techniques in student modeling. In *Proceedings of the 2nd World Summit on the Knowledge Society (WSKS 2009)*, Chania, Crete, Greece (pp. 552–564).

Durrani, S., & Durrani, D. S. (2010). Intelligent tutoring systems and cognitive abilities. In *Proceedings of Graduate Colloquium on Computer Sciences (GCCS), Department of Computer Science* (p. 1). FAST-NU Lahore.

Echauz, J. R., & Vachtsevanos, G. J. (1995). Fuzzy grading system. *IEEE Transactions on Education, 38*(2), 158–165.

Faraco, R. A., Rosatelli, M. C., & Gauthier, F. A. O. (2004). An approach of student modeling in a learning companion system. In *Proceedings of the IX IBERAMIA*, Puebla, Mexico (pp. 891–900).

Felder, R. M., & Silverman, L. K. (1988). Learning and teaching styles. *Engineering Education, 78*(7), 674–681.

Felder, R. M., & Soloman, B. A. (2003). Learning styles and strategies. Retrieved June 28, 2012, from http://www.ncsu.edu/felder-public/ILSdir/styles.htm.

Flavell, J. H. (1976) Metacognitive aspects of problem solving. In L. B. Resnick (Ed.), *The nature of intelligence* (pp 231–236). Hillsdale: Erlbaum.

Gaudioso, E., Montero, M., & Hernandez-del-Olmo, F. (2012). Supporting teachers in adaptive educational system through predictive models: A proof of concept. *Expert Systems with Applications, 39*(1), 621–625.

Gena, C. (2005). Methods and techniques for the evaluation of user-adaptive systems. *The knowledge Engineering Review, 20*(1), 1–37.

Gena, C., & Weibelzahl, S. (2007). Usability engineering for the adaptive web. In P. Brusilovsky, P. A. Kobsa, & W. Nejdl (Eds.), *The adaptive web: Methods and strategies of web personalization* (pp. 720–762). New York: Springer.

Geng, X., Qin, S.,Chang, H., & Yang, Y. (2011). A hybrid knowledge representation for the domain model of intelligent flight trainer. In *Proceedings 2011 IEEE International Conference on Cloud Computing and Intelligence Systems*, Beijing, China (pp. 29–33).

Glaser, R., Lesgold, A., & Lajoie, S. (1987). Toward a cognitive theory for the measurement of achievement. In R. Ronning, J. Glover, J. C. Conoley, & J. Witt (Eds.), *The influence of cognitive psychology on testing and measurement: The Buros-Nebraska symposium on measurement and testing* (pp. 41–85). Hillsdale, NJ: Erlbaum.

Glushkova, T. (2008). Adaptive model for user knowledge in the e-learning system. In *Proceedings of the International Conference on Computer Systems and Technologies* (CompSysTech'08).

Goel, G., Lallé, S., & Luengo, V. (2012). Fuzzy logic representation for student modelling case study on geometry. In *Proceedings of the 11th International Conference on Intelligent Tutoring Systems*, Chania, Greece (pp. 428–433).

Goguadze, G., Sosnovsky, S. A., Isotani, S., & McLaren, B. M. (2011a). Evaluating a bayesian student model of decimal misconceptions. In *Proceedings of the 4th International Conference on Educational Data Mining*, Eindhoven, the Netherlands (pp. 301–306).

Goguadze, G., Sosnovsky, S., Isotani, S., & McLaren, B. M. (2011b). Towards a bayesian student model for detecting decimal misconceptions. In *Proceedings 19th International Conference on Computers in Education*, Chiang Mai, Thailand (pp. 34–41).

Graesser, A. C., Conley, M. W., & Olney, A. M. (2012). Intelligent tutoring systems. In S. Graham & K. Harris (Eds.), *APA educational psychology handbook* (Vol. 3, pp. 451–473), Applications to learning and teaching Washington, DC: American Psychological Association.

Grigoriadou, M., Kornilakis, H., Papanikolaou, K. A., & Magoulas, G. D. (2002). Fuzzy inference for student diagnosis in adaptive educational hypermedia. In *Proceedings of the 2nd Hellenic Conference on AI: Methods and Applications of Artificial Intelligence*, Thessaloniki, Greece (pp. 191–202).

Grubišić, A., Stankov, S., Rosić, M., & Žitko, B. (2009). Controlled experiment replication in evaluation of e-learning system's educational influence. *Computers and Education,53*(3), 591–602.

Grubišić, A., Stankov, S., and Žitko, B. (2013). Stereotype Student Model for an Adaptive e-Learning System. *World Academy of Science, Engineering and Technology*, 7, 16–23.

Hernández, Y., Sucar, L., & Arroyo-Figueroa, G. (2010). Evaluating an affective student model for intelligent learning environments. In *Proceedings of the 12th Ibero-American Conference on Advances in Artificial Intelligence (IBERAMIA'10)*, Bahía Blanca, Argentina (pp. 473–482).

Holland, J., Mitrovic, A., & Martin, B. (2009). J-Latte: A constraint-based tutor for java. In *Proceedings of the 17th International Conference on Computers in Education*, Hong Kong (pp. 142–146).

Homsi, M., Lutfi, R., Rosa, M. C., & Bakarat, G. (2008). Student modeling using NN-HMM for EFL course. In *Proceedings of the 3rd International Conference on Information and Communication Technologies: From Theory to Applications (ICTTA 2008)*, Umayyad Palace, Damascus, Syria (pp. 1–6).

Idris, N., Yusof, N., & Saad, P. (2009). Adaptive course sequencing for personalization of learning path using neural network. *International Journal of Advances in Soft Computing and Its Applications, 1*(1), 49–61.

Inventado, P. S., Legaspi, R., Bui, T. D. & Suarez, M. (2010). Predicting student's appraisal of feedback in an its using previous affective states and continuous affect labels from EEG data. In *Proceedings of the 18th International Conference on Computers in Education*, Putrajaya, Malaysia (pp. 71–75).

Jameson, A. (1996). Numerical uncertainty management in user and student modeling: An overview of systems and issues. *User Modeling and User-Adapted Interaction, 5*(3–4), 193–251.

Jaques, N., Conati, C., Harley, J., & Azevedo, R. (2014). Predicting Affect from Gaze Data During Interaction with an Intelligent Tutoring System. In Proc. of the 12th International on Intelligent Tutoring Systems (ITS 2014), Honolulu, HI, USA, (pp 29–38).

Jarusek, P., & Pelánek, R. (2012). Modeling and predicting students problem solving times. In *Proceedings of the 38th International Conference on Current Trends in Theory and Practice of Computer Science (SOFSEM 2012)* (pp. 637–648).

Jeremić, Z., Jovanović, J., & Gasěvić, D. (2009). Evaluating an intelligent tutoring system for design patterns: The DEPTHS experience. *Educational Technology and Society,12*(2), 111–130.

Jeremić, Z., Jovanović, J., & Gasěvić, D. (2012). Student modeling and assessment in intelligent tutoring of software patterns. *Expert Systems with Applications,39*(1), 210–222.

Jia, B., Zhong, S., Zheng, T., & Liu, Z. (2010). The study and design of adaptive learning system based on fuzzy set theory. In Z. A. D. Cheok, W. Müller, X. Zhang, & K. Wong (Eds.), *Transactions on edutainment IV* (pp. 1–11). Berlin, Heidelberg: Springer.

Jili, C., Kebin, H., Feng, W., & Huixia, W. (2009). E-learning behavior analysis based on fuzzy clustering. In *Proceedings Third International Conference on Genetic and Evolutionary Computing (WGEC'09)* (pp. 863–866).

Johnson, W. L., Rickel, J. W., & Lester, J. C. (2000). Animated pedagogical agents: Face-to-face interaction in interactive learning environment. *International Journal of Artificial Intelligence in Education, 11*, 47–78.

Jonassen, D. H., & Grabowski, B. L. (1993). *Handbook of individual differences, learning and instruction.* Hillsdale, NJ: Lawrence Erlbaum Associates.

Jurado, F., Santos, O. C., Redondo, M. A., Boticario, J. G., & Ortega, M. (2008). Providing dynamic instructional adaptation in programming learning. In *Proceedings of the 3rd international workshop on Hybrid Artificial Intelligence Systems*, Burgos, Spain (pp. 329–336).

Karnik, N. N., Mendel, J. M., & Liang, Q. (1999). Type-2 fuzzy logic systems. *IEEE Transactions on Fuzzy Systems, 7*(6), 643–658.

Kass, R. (1991). Building a user model implicitly from a cooperative advisory dialog. *User Modeling and User-Adapted Interaction,1*, 203–258.

Kassim, A. A., Kazi, S. A., & Ranganath, S. (2004). A Web-based intelligent learning environment for digital systems. *International Journal of Engineering Education, 20*(1), 13–23.

Katsionis, G., & Virvou, M. (2004). A cognitive theory for affective user modelling in a virtual reality educational game. In *Proceedings of the IEEE International Conference on Systems, Man, and Cybernetics* (pp. 1209–1213).

Kavčič, A. (2004a). Fuzzy student model in intermediactor platform. In *Proceedings of the 26th International Conference on Information Technology Interfaces*, Croatia (pp. 297–302).

Kavčič, A. (2004b). Fuzzy user modeling for adaptation in educational hypermedia. *IEEE Transactions on Systems, Man and Cybernetics. Part C: Applications and Reviews, 34*(4), 439–449.

Kay, J. (2000). Stereotypes, student models and scrutability. In *Proceedings of the 5th International Conference on Intelligent Tutoring Systems*, Montréal, Canada (pp. 19–30).

Khamis, M. (2011). IDEAL: An intelligent distributed experience-based adaptive learning model. *Journal of Arts and Humanities, 20*(1).

Kirkpatrick, D. L. (1979). Techniques for evaluating training programs. *Training and Development Journal, 33*(6), 78–92.

Kofod-Petersen, A., Petersen, S. A., Bye, G. G., Kolås, L., & Staupe, A. (2008). Learning in an ambient intelligent environment—towards modeling learners through stereotypes. *Revue d'Intelligence Artificielle, 22*(5), 569–588.

Kosba, E., Dimitrova, V., & Boyle, R. (2003). Using fuzzy techniques to model students in web-based learning environment. In *Proceedings of the 7th International Conference on Knowledge-Based Intelligent Information and Engineering Systems*, United Kingdom (pp. 222–229).

Kosba, E., Dimitrova, V., & Boyle, R. (2005). Using student and group models to support teachers in web-based distance education. In *Proceedings of the International Conference on User Modeling*, Edinburgh (pp. 124–133).

Kumar, A. (2006a). Using enhanced concept map for student modeling in programming tutors. In *Proceedings of the 19th International Florida Artificial Intelligence Research Society Conference*, Melbourne Beach (pp. 527–532).

Kumar, A. (2006b). A scalable solution for adaptive problem sequencing and its evaluation. In *Proceedings of the 4th International Conference on Adaptive Hypermedia and Adaptive Web-Based Systems (AH'2006)*, Dublin, Ireland (pp. 161–171).

Kyriacou, D. (2008). A scrutable user modelling infrastructure for enabling life-long user modelling. In *Proceedings of the 5th International Conference on Adaptive Hypermedia and Adaptive Web-Based Systems*, Hannover, Germany (pp. 421–425).

Latham, A., Crockett, K., & McLean, D. (2014). An adaptation algorithm for an intelligent natural language tutoring system. *Computers and Education, 71*, 97–110.

Lavie, T., Meyer, J., Beugler, K., & Coughlin, J. F. (2005). The evaluation of in-vehicle adaptive systems. In *Proceedings of the 4th Workshop on the Evaluation of Adaptive Systems*, Edinburgh, UK (pp. 9–18).

Law, C. K. (1996). Using fuzzy numbers in education grading system. *Fuzzy Sets and Systems, 83*(3), 311–323.

Le, N. T., & Menzel, W. (2009). Using weighted constraints to diagnose errors in logic programming-the case of an Ill-defined domain. *Journal on Artificial Intelligence in Education, 19*(2), 382–400.

Lehman, B., Matthews. M., D'Mello, S., & Person, N. (2008). What are you feeling? Investigating student affective states during expert human tutoring sessions. In *Proceedings of the 9th International Conference on Intelligent Tutoring Systems (ITS 2008)*, Montreal (pp. 50–59).

Leon, M., Napoles, G., Garcia, M., Bello, R., & Vanhoof, K. (2011). Two steps Individuals travel behavior through fuzzy cognitive maps pre-definition and learning. In *Proceedings of the 10th Mexican International Conference on Artificial Intelligence*. Puebla, Mexico.

Li, N., Cohen, W. W., Koedinger, K. R., & Matsuda, N. (2011). A machine learning approach for automatic student model discovery. In *Proceedings of Conference on Educational Data Mining (EDM 2011)*, Eindhoven, the Netherlands (pp. 31–40).

Liang, Q., & Mendel, J. M. (2000). Interval type-2 fuzzy logic systems: Theory and design. *IEEE Transactions on Fuzzy Systems, 8*(5), 535–550.

Liaw, S.-S., & Huang, H.-M. (2013). Perceived satisfaction, perceived usefulness and interactive learning environments as predictors to self-regulation in e-learning environments. *Computers and Education, 60*, 14–24.

Limongelli, C., Sciarrone, F., Temperini, M., & Vaste, G. (2009). Adaptive learning with the LS-plan system: A field evaluation. *IEEE Transactions on Learning Technologies, 2*(3), 203–215.

Lin, C. M. (2007). Combination study of fuzzy cognitive map. *International Journal of Energy and Environment, 1*(2), 65–69.

Liu, C. L. (2008). Using bayesian networks for student modeling. In R. M. Viccari, P. Augustin-Jaques, & R. Verdin (Eds.), *Agent-based tutoring systems by cognitive and affective modeling* (pp. 97–113). Hershey: Igi Global.

Liu, Z., & Wang, H. (2007). A modeling method based on bayesian networks in intelligent tutoring system. In *Proceedings of the 11th International Conference on Computer Supported Cooperative Work in Design*, Melbourne, Australia (pp. 967–972).

Lo, J.-J., Chan, Y.-C., & Yen, S. W. (2012). Designing an adaptive web-based learning system based on students' cognitive styles identified online. *Computers and Education, 58*, 209–222.

Lu, C. H., Ong, C. S., & Hsu, W. L. (2007). Using an ITS as an arithmetic assistant for teachers 3-year review. *Journal of Internet Technology, 8*(4), 389–398.

Lu, C. H., Wu, C. W., Wu, S. H., Chiou, G. F., & Hsu, W. L. (2005). Ontological support in modeling learners' problem solving process. *Educational Technology and Society, 8*(4), 64–74.

Ma, J., & Zhou, D. (2000). Fuzzy set approach to the assessment of student-centered learning. *IEEE Transactions on Education, 43*(2), 237–241.

Mahnane, L., Laskri, M. T., & Trigano, P. (2012). An adaptive hypermedia system integrating thinking style (AHS-TS): Model and experiment. *International Journal of Hybrid Information Technology, 5*(1), 11–28.

Markham, S., Ceddia, J., Sheard, J., Burvill, C., Weir, J., Field, B. et al. (2003). Applying agent technology to evaluation tasks in e-learning environments. In *Exploring Educational Technologies Conference* (pp. 16–17).

Martin, B. (1999). Constraint-based student modeling: Representing student knowledge. In *Proceedings of the 3rd New Zealand Computer Science Research Students' Conference*, Hamilton NZ (pp. 22–29).

Martins, A. C., Faria, L., Vaz de Carvalho, C., & Carrapatoso, E. (2008). User modeling in adaptive hypermedia educational systems. *Educational Technology and Society, 11*(1), 194–207.

Mayo, M. J. (2001). Bayesian student modelling and decision-theoretic selection of tutorial actions in intelligent tutoring systems. *Ph.D. Thesis*. Retrieved June 29, 2012, from http://www.cosc.canterbury.ac.nz/research/reports/PhdTheses/2001/phd_0102.pdf.

McGill, T. J., & Volet, S. E. (1997). A conceptual framework for analyzing students' knowledge of programming. *Journal of Research on Computing in Education, 29*(3), 276–297.

Mendel, J. M. (2001). *Uncertain rule-based fuzzy logic systems: Introduction and new directions.* Upper Saddle River, NJ: Prentice-Hall.

Mendel, J. M. (2007). Advances in type-2 fuzzy sets and systems. *Information Sciences, 177*(1), 84–110.

Miao, Y., & Liu, Z. Q. (2000). On causal inference in fuzzy cognitive maps. *IEEE Transactions on Fuzzy Systems, 8*(1), 107–119.

Michaud, L. N., & McCoy, K. F. (2004). Empirical derivation of a sequence of user stereotypes for language learning. *User Modeling and User-Adapted Interaction, 14*, 317–350.

Millán, E., & Perez de la Cruz, J.-L. (2002). A bayesian diagnostic algorithm for student modeling. *User Modeling and User-Adapted Interaction, 12*(2/3), 281–330.

Millán, E., Loboda, T., & Pérez-de-la-Cruz, J. L. (2010). Bayesian networks for student model engineering. *Computers and Education, 55*(4), 1663–1683.

Mitrovic, A. (2003). An intelligent SQL tutor on the web. *International Journal of Artificial Intelligence in Education, 13*(2–4), 173–197.

Mitrovic, A., and Martin, B. (2006). Evaluating the effects of open student models on learning. In *Proceedings of the 2nd International Conference on Adaptive Hypermedia and Adaptive Web-Based Systems* (pp. 296–305).

Mitrovic, A., Marting, B., & Mayo, M. (2002). Using Evaluation to Shape ITS Design: Results and Experiences with SQL-Tutor. *User Modelling and User-Adapted Interaction, 12*(2/3), 243–279.

Mitrovic, A., Martin, B., & Suraweera, P. (2007). Intelligent tutors for all: Constraint-based approach. *IEEE Intelligent Systems, 22*(4), 38–45.

Mitrovic, A., Mayo, M., Suraweera, P., & Martin, B. (2001). Constraint-based tutors: A success story. In *Proceedings of 14th International Conference on Industrial and Engineering Applications of Artificial Intelligence and Expert Systems (IEA/AIE-2001)*, Budapest (pp. 931–940).

Mizumoto, M., & Tanaka, K. (1976). Some properties of fuzzy sets of type 2. *Information and Control, 31*(4), 312–340.

Moridis, C. N., & Economides, A. A. (2009). Prediction of student's mood during an online test using formula-based and neural network-based method. *Computers and Education, 53*, 644–652.

Mulwa, C., Lawless, S., Sharp, M., & Wade, V. (2011). The evaluation of adaptive and personalized information retrieval systems: A review. *International Journal of Knowledge and Web Intelligence, 2*(2/3), 138–156.

Muñoz, K., Mc Kevitt, P., Lunney, T., Noguez, J., & Neri, L. (2010). PlayPhysics: An emotional game learning environment for teaching physics. In *Proceedings of the 4th International Conference on Knowledge Science, Engineering and Management (KSEM' 10)*, Belfast, Northern Ireland, UK (pp. 400–411).

Muñoz, K., Mc Kevitt, P., Lunney, T., Noguez, J., & Neri, L. (2011). An emotional student model for game-play adaptation. *Entertainment Computing, 2*(2), 133–141.

Nguyen, L., & Do, P. (2008). Learner model in adaptive learning. In *Proceedings of World Academy of Science, Engineering and Technology* (pp. 396–401).

Nguyen, L., & Do, P. (2009). Combination of bayesian network and overlay model in user modeling. In *Proceedings of the 9th International Conference on Computational Science*, Baton Rouge, Louisiana, USA (pp. 5–14).

Nguyen, C. D., Vo, K. D., Bui, D. B., & Nguyen, D. T. (2011). An ontology-based IT student model in an educational social network. In *Proceedings of the 13th International Conference on Information Integration and Web-based Applications and Services (iiWAS '11)*, Bali, Indonesia (pp. 379–382).

Norusis, M. J. (2009). *SPSS 17.0 statistical procedures companion*. United States: Pearson Education.

Nwana, H. S. (1990). Intelligent tutoring systems: An overview. *Artificial Intelligence Review, 4*, 251–277.

Nykänen, O. (2006). Inducing fuzzy models for student classification. *Educational Technology and Society, 9*(2), 223–234.

Ohlsson, S. (1996). Learning from performance errors. *Psychological Review, 103*(2), 241–262.

Ortony, A., Clore, G. L., & Collins, A. (1988). *The cognitive structure of emotions*. Cambridge: Cambridge University Press.

Papageorgiou, E. (2011). Review study on fuzzy cognitive maps and their applications during the last decade. In *Proceedings of the IEEE International Conference on Fuzzy Systems. Taipei, Taiwan*.

Papageorgiou, E. I., & Iakovidis, D. K. (2013). Intuitionistic fuzzy cognitive maps. *IEEE Transactions on Fuzzy Systems, 21*(2), 342–354.

Papageorgiou, E. I., & Salmeron, J. L. (2012). Learning fuzzy grey cognitive maps using nonlinear hebbian-based approach. *International Journal of Approximate Reasoning,53*(1), 54–65.

Papanikolaou, K. A., Grigoriadou, M., Kornilakis, H., & Magoulas, G. D. (2003). Personalizing the interaction in a web-based educational hypermedia system: The case of INSPIRE. *User Modeling and User-Adapted Interaction, 13*(3), 213–267.

Pearl, J. (1988). *Probabilistic reasoning in expert systems: Networks of plausible inference*. San Francisco: Morgan Kaufmann Publishers Inc.

Pearl, J. (1996). Decision making under uncertainty. *ACM Computing Surveys, 28*(1), 89–92.

Pekrun, R., Frenzel, A. C., Goetz, T., & Perry, R. P. (2007). The control value theory of achievement emotions: An integrative approach to emotions in education. In P. A. Shutz & R. Pekrun (Eds.), *Emotion in education* (pp. 13–36). London: Elsevier.

Peylo, C., Teiken, W., Rollinger, C., & Gust, H. (2000). An ontology as domain model in a web-based educational system for prolog. In *Proceedings of the 13th International Florida Artificial Intelligence Research Society Conference*. Orlando, Florida, USA.

Parvez, S. M., & Blank, G. D. (2008). Individualizing tutoring with learning style based feedback. In *Proceedings of the 9th International Conference on Intelligent Tutoring Systems*, Montreal (pp. 291–301).

Peña, A., & Kayashima, M. (2011). Improving students' meta-cognitive skills within intelligent educational systems: A review. In *Proceedings of the 6th international conference on Foundations of Augmented Cognition: Directing the Future of Adaptive Systems* (pp. 442–451).

Peña, A., & Sossa, H. (2010). Semantic representation and management of student models: An approach to adapt lecture sequencing to enhance learning. In *Proceedings of the 9th Mexican International Conference on Advances in Artificial Intelligence: Part I*, Pachuca, Mexico (pp. 175–186).

Popescu, E. (2009). Diagnosing students' learning style in an educational hypermedia system. cognitive and emotional processes in web-based education: Integrating human factors and personalization. *Advances in Web-Based Learning Book Series* (pp. 187–208). US: IGI Global.

Popescu, E., Badica, C., & Moraret, L.(2009). WELSA: An intelligent and adaptive web-based educational system. In *Proceedings of the 3rd Symposium on Intelligent Distributed Computing*, Ayia Napa, Cyprus (pp. 175–185).

Popescu, E., Badica, C., & Moraret, L. (2010). Accommodating learning styles in an adaptive educational system. *Informatica, 34*, 451–462.

Pramitasari, L., Hidayanto, A. N., Aminah, S., Krisnadhi, A. A., & Ramadhanie, M. A. (2009). Development of student model ontology for personalization in an e-learning system based on semantic web. In *Proceedings of the International Conference on Advanced Computer Science and Information Systems (ICACSIS 2009)*, Universitas Indonesia, Depok, Indonesia (pp. 434–439).

Rich, E. (1979). User modelling via stereotypes. *Cognitive Science, 3*(4), 329–354.

Rivers, R. (1989). Embedded user models—where next? *Interacting with Computers, 1*, 14–30.

Rodrigo, M., Baker, R., Maria, L., Sheryl, L., Alexis, M., Sheila, P., Jerry, S., Leima, S., Jessica, S., & Sinath, T. (2007). Affect and usage choices in simulation problem solving environments. In *Proceedings of the 13th International Conference on Artificial Intelligence in Education*, Marina Del Ray, CA, USA (pp.145–152).

Rodriguez-Repiso, L., Setchi, R., & Salmeron, J. L. (2007). Modelling IT projects success with fuzzy cognitive maps. *Expert Systems with Applications, 32*(2), 543–559.

Sabourin, J., Mott, B., & Lester, J. C. (2011). Modeling learner affect with theoretically grounded dynamic bayesian networks. In *Proceedings of the 4th international conference on Affective computing and intelligent interaction*, Memphis, Tennessee (pp. 286–295).

Salim, N., & Haron, N. (2006). The construction of fuzzy set and fuzzy rulef or mixed approach in adaptive hypermedia learning system. In *Proceedings of the 1st International Conference on Technologies for E-Learning and Digital Entertainment (Edutainment 2006)*, Hangzhou, China (pp. 183–187).

Salmeron, J. L. (2009). Augment fuzzy cognitive maps for modelling LMS critical success factors. *Knowledge-Based Systems, 22*(4), 275–278.

Salmeron, J. L., Vidal, R., & Mena, A. (2012). Ranking fuzzy cognitive map based scenarios with TOPSIS. *Expert Systems with Applications, 39*(3), 2443–2450.

Salomon, G. (1990). Studying the flute and the orchestra: Controlled vs. classroom research on computers. *International Journal of Educational Research, 14*, 521–532.

Sison, R., & Shimura, M. (1998). Student modeling and machine learning. *International Journal of Artificial Intelligence in Education, 9*, 128–158.

Schiaffino, S., Garcia, P., & Amandi, A. (2008). eTeacher: Providing personalized assistance to e-learning students. *Computers and Education, 51*(4), 1744–1754.

Shakouri, H. G., & Menhaj, M. (2008). A systematic fuzzy decision-making process to choose the best model among a set of competing models. *IEEE Trans on Systems Man and Cybernetics Part A: Systems and Humans, 38*(5), 1118–1128.

Shapiro, J. A. (2005). An algebra subsystem for diagnosing students' input in a physics tutorin system. *International Journal of Artificial Intelligence in Education, 15*(3), 205–228.

Siddapa, M., & Manjunath, A. S. (2007). Knowledge representation using multilevel hierarchical model in intelligent tutoring system. In *Proceedings of the Third International Conference on Advances in Computer Science and Technology*. Thailand.

Siddappa, M., Manjunath, A. S., & Kurian, M. Z. (2009). Design, implementation and evaluation of intelligent tutoring system for numerical methods (ITNM). *International Conference on Computational Intelligence and Software Engineering*, Wuhan, China (pp. 1–7).

Smith, S. (1998). Tutorial on. Retrieved March 15, 2002, from http://www.cs.mdx.ac.uk/staffpages/serengul/table.of.contents.htm.

Song, H., Miao, C., Roel, W., Shen, Z., & D'Hondt, M. (2011). An extension to fuzzy cognitive maps for classification and prediction. *IEEE Transactions on Fuzzy Systems, 19*(1), 116–135.

Spada, H. (1993). How the role of cognitive modeling for computerized instruction is changing. In *Proceedings of AI-ED'93, World Conference on Artificial Intelligence in Education*, Edinburgh, Scotland (pp. 21–25).

Staff, C. (2001). HyperContext: A framework for adaptive and adaptable hypertext. *Ph.D. Thesis. University of Sussex.*

Stansfield, J. C., Carr, B., & Goldstein, I. P. (1976). Wumpus advisor I: A first implementation of a program that tutors logical and probabilistic reasoning skills. *At Lab Memo 381.* Cambridge, Massachusetts: Massachusetts Institute of Technology.

Stash, N., Cristea, A., & De Bra, P. (2006). Adaptation to Learning Styles in E-Learning: Approach Evaluation. In *Proc. of World Conference on E-Learning in Corporate, Government, Healthcare, and Higher Education*, Orlando, Florida (pp. 284–291).

Stathacopoulou, R., Magoulas, G. D., Grigoriadou, M., & Samarakou, M. (2005). Neuro-fuzzy knowledge processing in intelligent learning environments for improved student diagnosis. *Information Sciences, 170*(2–4), 273–307.

Stula, M., Stipanicev, D., & Bodrozic, L. (2010). Intelligent modeling with agent-based fuzzy cognitive map. *International Journal on Intelligent Systems, 25*(10), 981–1004.

Stylios, C. D., & Groumpos, P. P. (2004). Modeling complex systems using fuzzy cognitive maps. *IEEE Transactions on Systems Man and Cybernetics: Part A, 34*(1), 155–162.

Suarez-Cansino, J., & Hernandez-Gomez, A. (2008). Adaptive testing system modeled through fuzzy logic. In *Proceedings of the Second WSEAS International Conference on Computer Engineering and Applications (CEA'08)*, Acapulco, Mexico (pp. 85–89).

Sucar, L. E., & Noguez, J. (2008). Student modeling. In O. Pourret, P. Naom, & B. Marcot (Eds.), *Bayesian networks: A practical guide to applications* (pp. 173–185). West Sussex: Wiley.

Suraweera, P., & Mitrovic, A. (2004). An intelligent tutoring system for entity-relationship modelling. *Artificial Intelligence in Education, 14*(3–4), 375–417.

Surjono, H., & Maltby, J. (2003). Adaptive educational hypermedia based on multiple student characteristics. In *Proceedings of the 2nd International Conference on Web-based Learning*, Melbourne, Australia (pp. 442–449).

Thomson, D., & Mitrovic, A. (2009). Towards a negotiable student model for constraint-based ITSs. In *Proceedings of the 17th International Conference on Computers in Education*, Hong Kong (pp. 83–90).

Ting, C.-Y., & Phon-Amnuaisuk, S. (2012). Properties of Bayesian student model for INQPRO. *Applied Intelligence, 36*(2), 391–406.

Tourtoglou, K., & Virvou, M. (2008). User stereotypes concerning cognitive, personality and performance issues in a collaborative learning environment for UML. In *Proceedings of the 1st International Symposium on Intelligent Interactive Multimedia Systems and Services (KES-IIMSS 2008)*, Piraeus, Greece (pp. 385–394).

Tourtoglou, K., & Virvou, M. (2012). An intelligent recommender system for trainers and trainees in a collaborative learning environment for UML. *Journal of Intelligent Decision Technologies, 6*(2), 79–95.

Tretiakov, A., Sridharan, B., & Kinshuk, K. (2005). Conceptual modelling of web-based tutoring systems. In *Proceedings of the International Conference on Computers in Education*, Altona and Melbourne, Australia (pp. 2051–2058).

Tsaganoua, G., Grigoriadou, M., Cavoura, T., & Koutra, D. (2003). Evaluating an intelligent diagnosis system of historical text comprehension. *Expert Systems with Applications, 25*, 493–502.

Tsiriga, V., & Virvou, M. (2002). Initializing the student model using stereotypes and machine learning. In *Proceedings of the 2002 IEEE International Conference on System, Man and Cybernetics* (pp. 404–409).

Tsiriga, V., & Virvou, M. (2003a). Modelling the student to individualise tutoring in a web-based ICALL. *International Journal of Continuing Engineering Education and Lifelong Learning, 13*(3–4), 350–365.

Tsiriga, V., & Virvou, M. (2003b). Evaluation of an intelligent web-based language tutor. In *Proceedings of the 7th International Conference on Knowledge-Based Intelligent Information and Engineering Systems (KES 2003)*, Oxford, UK (pp. 275–281).

Tsiriga, V., & Virvou, M. (2003c). Initializing student models in web-based ITSs: A generic approach. In *Proceedings of the 3rd International Conference on Advanced Learning Technologies (ICALT 2003)*, Athens, Greece (pp. 42–46).

Vasandani, V., & Govindaraj, T. (1995). Knowledge organization in intelligent tutoring systems for diagnostic problem solving in complex dynamic domains. *IEEE Transactions on Systems, Man and Cybernetics, 25*(7), 1076–1096.

Vélez, J., Fabregat, R., Nassiff, S., Petro, J., & Fernandez, A. (2008). User integral model in adaptive virtual learning environment. In *Proceedings of World Conference on E-Learning in Corporate, Government, Healthcare, and Higher Education*, Las Vegas, Nevada, United States (pp. 3275–3284).

Viccari, R. M., Flores, C. D., Seixas, L., Gluz, J. C., & Coelho, H. (2008). AMPLIA: A probabilistic learning environment. *International Journal of Artificial Intelligence in Education, 18*(4), 347–373.

Virvou, M., & Kabassi, K. (2002). F-SMILE: An intelligent multi-agent learning environment. In *Proceedings of IEEE International Conference on Advanced Learning Technologies 2002 (ICALT'02)* (pp. 144–149).

Virvou, M., & Kabassi, K. (2004). Evaluating an intelligent graphical user interface by comparison with human experts. *Knowledge-Based Systems, 17*(1), 31–37.

Virvou, M., Katsionis, G., & Manos, K. (2005). Combining software games with education: Evaluation of its educational effectiveness. *Educational Technology and Society, 8*(2), 54–65.

Virvou, M., Lampropoulos, A. S. & Tsihrintzis, G. A. (2006). Inma: A knowledge-based authoring tool for music education. In *Proceedings of the 10th International Conference on Knowledge-Based Intelligent Information and Engineering Systems*, United Kingdom (pp. 376–383).

Wang, H. Y., & Chen, S. M. (2006). New methods for evaluating students' answerscripts using fuzzy numbers associated with degrees of confidence. In *Proceedings of the 2006 IEEE International Conference on Fuzzy Systems*, Vancouver, BC, Canada (pp. 5492–5497).

Wang., X., Yang, Y., & Wen, X. (2009). Study on blended learning approach for english teaching. In *Proceedings of the 2009 IEEE International Conference on Systems, Man, and Cybernetics* (pp. 4641–4644).

Weerasinghe, A., & Mitrovic, A. (2011). Facilitating adaptive tutorial dialogues in EER-tutor. In *Proceedings of the 15th International Conference on Artificial Intelligence in Education, Auckland*, New Zealand (pp. 630–631).

Webb, G. (1998). Preface to UMUAI special issue on machine learning for user modeling. *User Modeling and User-Adapted Interaction, 8*, 1–3.

Webb, G., Pazzani, M., & Billsus, D. (2001). Machine learning for user modeling. *User Modeling and User-Adapted Interaction, 11*, 19–29.

Weibelzahl, S., & Weber, G. (2003). Evaluation of the inference mechanism of adaptive learning systems. In *Proceedings of the 9th International Conference on User modeling*, Johnstown, PA, USA (pp. 154–168).

Welch, M., & Brownell, K. (2000). The development and evaluation of a multimedia course on educational collaboration. *Journal of Educational Multimedia and Hypermedia, 9*(3), 169–194.

Weon, S., & Kim, J. (2001). Learning achievement evaluation strategy using fuzzy membership function. In *Proceedings of the 31st ASEE/IEEE Frontiers In Education Conference*, Reno, NV (pp. 19–24).

Wilson, E., Karr, C. L., & Freeman, L. M. (1998). Flexible, adaptive, automatic fuzzy-based grade assigning system. In *Proceedings of the 1998 north American Fuzzy Information Processing Society (NAFIPS) Conference* (pp. 334–338).

Winter, M., Brooks, C., & Greer, J. (2005).Towards best practices for semantic web student modeling. In *Proceedings of the 12th International Conference on Artificial Intelligence in Education*, Amsterdam, the Netherlands (pp. 694–701).

Wisher, R. A., & Fletcher, J. D. (2004). The case for advanced distributed learning. *Information and Security: An International Journal, 14*, 17–25.

Xu, D., Wang, H., & Su, K. (2002). Intelligent student profiling with fuzzy models. In *Proceedings of the 35th Hawaii International Conference on System Sciences*.

Yang, G., Kinshuk, K., & Graf, S. (2010). A practical student model for a location-aware and context-sensitive personalized adaptive learning system. In *Proceedings of the IEEE Technology for Education Conference, Bombay*, India (pp. 130–133).

Zadeh, L. A. (1965). Fuzzy sets. *Information and Control, 8*(3), 338–353.

Zadeh, L. A. (1975). Fuzzy logic and approximate reasoning. *Synthese, 30*, 407–428.

Zadeh, L. A. (1979). A theory of approximate reasoning. In J. Hayes, D. Michie, & L. I. Mikulich (Eds.), *Machine Intelligence 9* (pp. 149–194). New York: Halstead Press.

Zadeh, L. A. (2001). A new direction in AI—toward a computational theory of perceptions. *AI Magazine, 22*(1), 73–84.

Zapata-Rivera, D. (2007). Indirectly visible Bayesian student models. In *Proceedings of the 5th UAI Bayesian Modelling Applications Workshop*. Vancouver, BC, Canada.

Zatarain-Cabada, R., Barrón-Estrada M. L., Angulo V. P., García, A. J., & García C. A. R. (2010). Identification of felder-silverman learning styles with a supervised neural network. In *Proceedings of the 6th International Conference on Intelligent Computing (ICIC 2010)*, Changsha, China (pp. 479–486).

Printed in the United States
By Bookmasters